NATURE AND NURTURE IN FRENCH SOCIAL SCIENCES, 1859–1914 AND BEYOND

McGill-Queen's Studies in the History of Ideas
Series Editor: Philip J. Cercone

1 Problems of Cartesianism
Edited by Thomas M. Lennon, John M. Nicholas, and John W. Davis

2 The Development of the Idea of History in Antiquity
Gerald A. Press

3 Claude Buffier and Thomas Reid: Two Common-Sense Philosophers
Louise Marcil-Lacoste

4 Schiller, Hegel, and Marx: State, Society, and the Aesthetic Ideal of Ancient Greece
Philip J. Kain

5 John Case and Aristotelianism in Renaissance England
Charles B. Schmitt

6 Beyond Liberty and Property: The Process of Self-Recognition in Eighteenth-Century Political Thought
J.A.W. Gunn

7 John Toland: His Methods, Manners, and Mind
Stephen H. Daniel

8 Coleridge and the Inspired Word
Anthony John Harding

9 The Jena System, 1804–5: Logic and Metaphysics
G.W.F. Hegel
Translation edited by John W. Burbidge and George di Giovanni
Introduction and notes by H.S. Harris

10 Consent, Coercion, and Limit: The Medieval Origins of Parliamentary Democracy
Arthur P. Monahan

11 Scottish Common Sense in Germany, 1768–1800: A Contribution to the History of Critical Philosophy
Manfred Kuehn

12 Paine and Cobbett: The Transatlantic Connection
David A. Wilson

13 Descartes and the Enlightenment
Peter A. Schouls

14 Greek Scepticism: Anti-Realist Trends in Ancient Thought
Leo Groarke

15 The Irony of Theology and the Nature of Religious Thought
Donald Wiebe

16 Form and Transformation: A Study in the Philosophy of Plotinus
Frederic M. Schroeder

17 From Personal Duties towards Personal Rights: Late Medieval and Early Modern Political Thought, 1300–1650
Arthur P. Monahan

18 The Main Philosophical Writings and the Novel *Allwill*
Friedrich Heinrich Jacobi
Translated and edited by George di Giovanni

19 Kierkegaard as Humanist: Discovering My Self
Arnold B. Come

20 Durkheim, Morals, and Modernity
W. Watts Miller

21 The Career of Toleration: John Locke, Jonas Proast, and After
Richard Vernon

22 Dialectic of Love: Platonism in Schiller's Aesthetics
David Pugh

23 History and Memory in Ancient Greece
Gordon Shrimpton

24 Kierkegaard as Theologian: Recovering My Self
Arnold B. Come

25 Enlightenment and Conservatism in Victorian Scotland: The Career of Sir Archibald Alison
Michael Michie

26 The Road to Egdon Heath: The Aesthetics of the Great in Nature
Richard Bevis

27 Jena Romanticism and Its Appropriation of Jakob Böhme: Theosophy – Hagiography – Literature
Paolo Mayer

28 Enlightenment and Community: Lessing, Abbt, Herder, and the Quest for a German Public
Benjamin W. Redekop

29 Jacob Burckhardt and the Crisis of Modernity
John R. Hinde

30 The Distant Relation: Time and Identity in Spanish-American Fiction
Eoin S. Thomson

31 Mr Simson's Knotty Case: Divinity, Politics, and Due Process in Early Eighteenth-Century Scotland
Anne Skoczylas

32 Orthodoxy and Enlightenment: George Campbell in the Eighteenth Century
Jeffrey M. Suderman

33 Contemplation and Incarnation: The Theology of Marie-Dominique Chenu
Christophe F. Potworowski

34 Democratic Legitimacy: Plural Values and Political Power
F.M. Barnard

35 Herder on Nationality, Humanity, and History
F.M. Barnard

36 Labeling People: French Scholars on Society, Race, and Empire, 1815–1849
Martin S. Staum

37 The Subaltern Appeal to Experience: Self-Identity, Late Modernity, and the Politics of Immediacy
Craig Ireland

38 The Invention of Journalism Ethics: The Path to Objectivity and Beyond
Stephen J.A. Ward

39 The Recovery of Wonder: The New Freedom and the Asceticism of Power
Kenneth L. Schmitz

40 Reason and Self-Enactment in History and Politics: Themes and Voices of Modernity
F.M. Barnard

41 The More Moderate Side of Joseph de Maistre: Views on Political Liberty and Political Economy
Cara Camcastle

42 Democratic Society and Human Needs
Jeff Noonan

43 The Circle of Rights Expands: Modern Political Thought after the Reformation, 1521 (Luther) to 1762 (Rousseau)
Arthur P. Monahan

44 The Canadian Founding: John Locke and Parliament
Janet Ajzenstat

45 Finding Freedom: Hegel's Philosophy and the Emancipation of Women
Sara MacDonald

46 When the French Tried to Be British: Party, Opposition, and the Quest for the Civil Disagreement, 1814–1848
J.A.W. Gunn

47 Under Conrad's Eyes: The Novel as Criticism
Michael John DiSanto

48 Media, Memory, and the First World War
David Williams

49 An Aristotelian Account of Induction: Creating Something from Nothing
Louis Groarke

50 Social and Political Bonds: A Mosaic of Contrast and Convergence
F.M. Barnard

51 Archives and the Event of God: The Impact of Michel Foucault on Philosophical Theology
David Galston

52 Between the Queen and the Cabby: Olympe de Gouges's *Rights of Women*
John R. Cole

53 Nature and Nurture in French Social Sciences, 1859–1914 and Beyond
Martin S. Staum

NATURE AND NURTURE IN FRENCH SOCIAL SCIENCES, 1859–1914 AND BEYOND

Martin S. Staum

McGill-Queen's University Press
Montreal & Kingston • London • Ithaca

ISBN 978-0-7735-3892-4

Legal deposit third quarter 2011
Bibliothèque nationale du Québec

Printed in Canada on acid-free paper that is 100% ancient forest free (100% post-consumer recycled), processed chlorine free

This book has been published with the help of a grant from the Canadian Federation for the Humanities and Social Sciences, through the Aid to Scholarly Publications Programme, using funds provided by the Social Sciences and Humanities Research Council of Canada.

McGill-Queen's University Press acknowledges the support of the Canada Council for the Arts for our publishing program. We also acknowledge the financial support of the Government of Canada through the Canada Book Fund for our publishing activities.

Library and Archives Canada Cataloguing in Publication

Staum, Martin S., 1943–
Nature and nurture in French social sciences, 1859–1914 and beyond / Martin S. Staum.

(McGill-Queen's studies in the history of ideas; 53)
Includes bibliographical references and index.
ISBN 978-0-7735-3892-4

1. Social sciences–France–History–19th century. 2. Social sciences–France–History–20th century. 3. Nature and nurture–France–History–19th century. 4. Nature and nurture–France–History–20th century. I. Title. II. Series: McGill-Queen's studies in the history of ideas; 53

H53.F7S73 2011 300'94409034 C2011-902773-9

This book was typeset by Interscript in 10/12 New Baskerville.

To Sarah, Nina, Wally, Libby, and Elliot

Contents

Acknowledgments xi

Illustrations xiii

1 Introduction to the Nature-Nurture Debate 3

2 The Ethnographers 17

3 The Anthropologists 46

4 Ribot and Psychological Heredity 84

5 Heredity and Milieu in the *Revue philosophique* 108

6 Alfred Binet: From Measuring Heads to Testing Intelligence 126

7 The Non-Durkheimian Sociologists 142

8 Reorientation of Institutions 169

9 Echoes in the Vichy Era 194

10 Conclusion 205

Bibliography 213

Index 253

Acknowledgments

AT THE OUTSET OF THIS PROJECT, the French historian of anthropology Claude Blanckaert assured me that nineteenth-century anthropologists were all believers in the influence of heredity and that they were also all believers in the influence of the environment. Little did I realize at the time the wisdom of this indication that nature and nurture could never be completely opposed, for one of the major themes in this study is how the concepts of the innate and acquired are inextricably bound up with each other. For his acute perceptions and his profound knowledge of the history of the social sciences, I should like to thank Claude Blanckaert. For very helpful advice, encouragement, and hospitality, I am grateful to the historian of psychology Jacqueline Carroy. I have also benefited from the insights of Laurent Mucchielli, who encouraged me in the early stages of my research, though he may not agree with my conclusions.

For recent work that confirms many of my original assumptions, I would like to acknowledge Carole Reynaud Paligot, whom I know only as a correspondent. To Alain Froment of the Société d'anthropologie de Paris, I should like to extend thanks for allowing consultation of society archives, with the gracious assistance of Frédérique Valentin and Pascal Sellier. To the many librarians and archivists who have facilitated my research, I am especially thankful for your consideration. I am also obliged to a Master's student, Katie Emond, for research assistance. I would also like to acknowledge the very helpful advice and diligence of McGill-Queen's editor Joan McGilvray and copy editor Lesley Andrassy. Of course, no one named here is responsible for any errors of fact or interpretation in this study.

I very much appreciate the funding from the Social Sciences and Humanities Research Council of Canada that made possible several research expeditions to France.

Some ideas for chapters 2 and 3 of this study appeared in a preliminary form in "Nature and Nurture in French Ethnography and Anthropology, 1859–1914," *Journal of the History of Ideas* 65, 3 (2004): 475–95. Portions of chapters 4 and 6 appeared first in "Ribot, Binet, and the Emergence from the Anthropological Shadow," *Journal of the History of the Behavioral Sciences,* 43, 1 (2007): 1–18 and portions of chapter 7 in "'Race' and Gender in Non-Durkheimian French Sociology, 1893–1914," *Canadian Journal of History* 42, 2 (2007): 183–208.

There is no way I can sufficiently acknowledge the invaluable support of my wife Sarah for more than half a lifetime.

Seal of the Société d'ethnographie de Paris, *L'Ethnographie*, nouvelle série, 1 (1913) title page, Département des Imprimés, Bibliothèque nationale de France

Léon de Rosny, Founder of the Société d'ethnographie de Paris, from D. Marceron, ed. *Pensées extraites des ouvrages de Léon de Rosny*, Cullman Library, Smithsonian Institution

Paul Broca, Founder of the Société d'anthropologie de Paris. Collection of the National Library of Medicine

Charles Letourneau, Sociological Anthropologist, Département des Estampes et de Photographie, Bibliothèque nationale de France

Théodule Ribot, Psychologist and Philosopher. Département des Estampes et de Photographie, Bibliothèque nationale de France

NATURE AND NURTURE IN FRENCH SOCIAL SCIENCES, 1859–1914 AND BEYOND

1

Introduction to the Nature-Nurture Debate

EVEN AS GENETICISTS probe ever more deeply the secrets of the human genome, the perennial debate continues over the influence of hereditary nature versus cultural and environmental nurture on human behaviour. In the most profound sense, the debate has ended in a draw. For even the most enthusiastic geneticist admits that expression of genes is not automatic and varies with circumstances. Even the most ardent critic of biological determinism would concede that the evolutionary adaptation of the human brain enables the formation of human culture. But as ideological disputes continue in the twenty-first century about just how much humans can alter their destiny, the entire furor concerns the weight assignable to each side and whether or how they are inextricably intertwined. At stake is the limiting constraint of a cross-cultural human nature against those who argue that parenting (as well as the peer group) makes enormous differences in the lives of children, that genetic differences between men and women do not pre-program their patterns of life or career preferences, that education or example can counteract drives of aggression, and that treatment as well as punishment may affect the behaviour of criminals.

Almost every era has sufficient social turmoil to make these issues timely. The great debates in nineteenth- and early twentieth-century France illuminate our current dilemmas. In late nineteenth-century France, a host of social and political fears assailed the national psyche – relentless urbanization and industrialization, the involvement of the masses in politics, the fear of national decline after the loss of the Franco-Prussian war of 1870–71 amid low birth rates and possible "degeneration," and forebodings about national identity because of immigration and mixing of ethnic groups.[1] The emerging human sciences in mid-nineteenth-century France held high stakes in this controversy – the educability of individuals

1 Nye, *Crime, Madness, and Politics*; on immigration see Noiriel, *Immigration, antisémitisme, et racisme.*

within a society, including the prospects for women, the identity and significance of groups making up a nationality, and the place in a so-called "ladder of being" of all the peoples of the earth, with all its possible consequences for the quest for empire. While issues of crime and immigration have recently reappeared in France in more distressing form, the great difference in assumptions then from the theories of the most determined geneticists now was the still routine belief in inferiority of the feminine intellect, and the alleged moral, psychological, and cultural retardation associated with the physical racial characteristics of vast non-European populations.

Despite a seemingly glaring contradiction between the rights of republican citizens and the justification of domination of "inferiors," some scholars have argued that the very nature of defining citizenship in a republic was exclusionary – to women and ultimately to people of colour. In this view abstract republican universalism could not tolerate the reality of "difference."[2] But there may be more deeply rooted cultural prejudices than republican universalism. It was not only under republics that concepts of racial and gender superiority flourished. The naturalists' impulse to classify and the practices of slavery had affected the fledgling human sciences as early as the mid-eighteenth century and certainly long before the consolidation of the Third Republic, which is considered the golden age of "scientific racism."[3] One of the most determined efforts in France to study physical and cultural human attributes was the work of the Société ethnologique de Paris from 1839 to 1848. During the monarchy of Louis-Philippe they had already endorsed the ideology of a "civilizing mission" to racial or cultural inferiors that justified imperial rule.[4] Moreover, conservative thinkers such as Gustave Le Bon, uncomfortable with the very idea of a republic or mass democracy, were also extremely intolerant of women in the public sphere.[5]

Late nineteenth-century discussions of heredity and milieu, as in similar debates today, rarely supported polar extremes. Most often their conclusions reluctantly admitted the import of both nature and nurture.

2 Wilder, *The French Imperial Nation-State*, on the "imperial nation-state" as a single entity, so that there is no contradiction between the Declaration of Rights of Man and imperial rule; on gender issues, Landes, *Women and the Public Sphere*; for more recent arguments on universalism, Scott, *Parité;* for the equivalence of marginalization of women and "inferior races," Stepan, "Race, Gender, Science, and Citizenship," 26–52.

3 Dorlin, *La Matrice de la race.*

4 On the Société d'ethnologique de Paris, Blanckaert, "On the Origins of French Ethnology,"18–55; Staum, "Paris Ethnologists and the Perfectibility of 'Races,'" 453–72; on the civilizing mission see Conklin, *A Mission to Civilize.*

5 Nye, *The Origins of Crowd Psychology*, 46–50 on the gullible feminine mind and the crowd.

Most hereditarians made some concessions to the partisans of the milieu, and most believers in climatic or social influences did not ignore the legacy of physical inheritance. Yet the relative balance was all-important. Often greater emphasis on heredity led to greater belief in potential limitations on the intellect and character of "inferiors." This study aims to show first how the reputed influence of heredity coalesced into theories of race and gender affecting the image of non-Europeans and women. But important insights also run counter to the conventional wisdom. The advocates of environmental influence might not easily turn into believers in egalitarianism.

Ideological presuppositions might have influenced the authors, but no one-to-one correspondence existed between hereditarians and conservatives or progressives and belief in the greater effect of milieu. Certainly the natural inclination of conservatives was to promote social stability and doubt the malleability of "human nature." Pioneering researchers on "hereditary genius," such as Darwin's independently wealthy cousin Francis Galton, did in fact overstress heredity while subscribing to racial and social elitism as well as masculine superiority.[6] Reformers or even relatively moderate French republicans highlighted the power of the milieu because they wished to remake human nature, or to deny any cross-cultural human nature in favour of a socially and culturally embedded humanity.

Diane Paul and Michael Freeden have already shown the fallacy of any overarching assumption about the conservative-hereditarian and the progressive-environmental axes. There could be ideological affinities between socialism and the hereditarian desire to improve the species (Galton's "eugenics"), though often eugenics in France included environmental assumptions. As in the work of Carole Paligot, we shall find here that staunch French republicans and socially progressive thinkers were in fact quite eager to rank human beings hierarchically and to assume a strong hereditary limit on human capability. Whether or not such views were inherent in republican universalism, republicans had no compunction about endorsing domination of others or accepting the civilizing mission of tutoring others as childlike inferiors.[7] Conversely, a traditional or even deist religious outlook could sometimes be a force for belief in human brotherhood before God. Other staunch Christians

6 Cowan, *Sir Francis Galton*; Cowan, "Nature and Nurture," 133–208; see also the warnings of the exigencies of statistics in setting the postulates of Galton in Waller, "Putting Method First," 35–62 and Sweeney, *Fighting for the Good Cause*.

7 Freeden, "Eugenics and Progressive Thought: A Study in Ideological Affinity," 645–7; Paul, *The Politics of Heredity*, "Eugenics and the Left," 11–36 and the "Debate that Never Dies" 81–93; Lewontin, "Race and Intelligence," 5 on genetic factors of some importance; Paligot, *Races, racisme et antiracisme*.

remained supporters of the imperial imperative and believers in a racial hierarchy. In either case, however, ideological constraints affected the nature-nurture debate and the formation of the social sciences.

Consideration of two learned societies founded in 1859, the same year Charles Darwin published the *Origin of Species*, enhances our understanding of the complexity of the nature-nurture problem. The pioneering work of Elizabeth Williams has portrayed the Société d'ethnographie américaine et orientale as a group of literary scholars enamoured of ancient and current exotic civilizations and sensitive to cultural influences on peoples. On the other hand, the Société d'anthropologie de Paris was largely famed for its obsession with physical measurement, especially of heads and crania, in the hope of relating hereditary physical appearance to intellect and character.

Chapter 2 shows that the relatively little studied ethnographers had no uniform outlook on the nature-nurture question. As a group of linguists, geographers, and students of literature, with a leavening of wealthy or aristocratic members, they anticipated some aspects of cultural anthropology, though they studied texts without fieldwork. Their tolerance for other cultures impelled this society to proclaim itself the harbinger of worldwide fraternity. But the influence of nurture did not eclipse an indelible nature. For through the rhetoric of their programmatic egalitarian statements seeped a disarming fascination for physical features and telltale expressions of racial superiority.[8]

An abundant literature exists about the anthropologists who are the subject of chapter 3. Among a plethora of articles, Claude Blanckaert's essay on the "crisis of anthropometry" clarifies some of the direct political implications of the anthropologists' efforts, including their "expertise" on minimum height requirements for the French army. He also describes the "crisis" of difficulty of consistent measurements for racial distinctions – a problem that affected all the social sciences in the late nineteenth century. He has recently revised some of his articles and added substantially new material in an encyclopedic study of Broca and the Société d'anthropologie de Paris – a work that appeared after this study was completed. His theme of debates among anthropologists, without a unified voice on several important issues, agrees with the view of factions in the society that is presented in the third chapter.[9] Jennifer Hecht, without stressing racial hierarchy, has discussed the society as an anti-clerical, secularist, progressive republican coterie, predominantly of physicians, that

8 For a description of the society, Robert Lacombe, "Essai sur les origines," 329–41; Chailleu, "Histoire de la Société d'ethnographie," 89–107.

9 Blanckaert, "La Crise de l'Anthropométrie," 95–172; Blanckaert, *De la Race.*

in some cases became active in political opposition circles during the Empire of Napoleon III.[10] More recently than Elizabeth Williams, Jean Wartelle has studied the institutional functioning of the society.[11] Yet the literature still needs an analysis of the wide spectrum of anthropologists' views. Some were hard-line partisans of physical heredity while others admitted the importance of the climatic, the sociological, and the cultural. Their most distinctive interpretation of Lamarckian inheritance of acquired characteristics led them to build up the initial influence of the milieu only to nullify its impact in later eras.

The background to the creation of the anthropology society and the attempt to create a scientific study of humankind arose in part out of the long-standing medical tradition of relating physical to mental temperament and of painstaking consideration of human anatomy and physiology.[12] But the leading motivation of Paul Broca (1824–1880), the founder of the Société d'anthropologie de Paris, was the classificatory imperative of natural history. For Broca wished to revise the concept of animal hybridity to argue for the possibility of fertile hybrids of distinct species. The implication was that fertile interracial mixing in humans (the degree of fertility was still doubted and graded according to the proximity of the races) did not prove the unity of the human species. The purpose was not only to place human beings in nature, but also to show the plausibility of several distinct human species while mapping the relationships and capabilities of all human "races."[13]

Many anthropologists accepted the Lamarckian view that organisms adapted to physical environments and then inherited acquired characteristics. Despite their interest in physical measurements, they could thus not completely privilege heredity and ignore milieu. The controversial question was whether the once powerful geographic milieu or the continually powerful social milieu could still induce significant changes in intellect and character. Almost all who acknowledged the power of the milieu at that time set limits to its potential effects on peoples lacking "perfectibility." Others reasoned that the milieu could most notably affect those with superior racial attributes or advanced culture, including European women, but not most "retarded" or innately less endowed non-Europeans. The anthropologists for the most part illustrate the inextricable association of belief in heredity and belief in the milieu, but often with qualifications that made the milieu less powerful. A minority

10 Hecht, *The End of the Soul.*

11 Wartelle, "La Société d'Anthropologie de Paris," 125–71.

12 Williams, *The Physical and the Moral*; Staum, *Cabanis*; Blanckaert, "La Création de la chaire d'anthropologie," 85–123.

13 Schiller, *Paul Broca*, 129–31; Blanckaert, "Préface," i-xliii.

countercurrent within the Paris society, while not giving up brain or head measurement, valued the influence of geographic, social, and cultural surroundings upon humans. Thus the anthropologists in some ways turned out to be a mirror image of the ethnographers – refusing to give up the cultural domain – while the ethnographers also managed to cling to the physical.[14]

In recent studies, Carole Paligot, building on the work of Claude Blanckaert, has convincingly argued for a "racial paradigm" that affected all of the social sciences.[15] My interpretation of the subject matter of chapters 4 to 7 endorses this argument for a cross-disciplinary perspective to understand the use of a hard-line hereditarian anthropological approach in the emerging disciplines of psychology and sociology. Aside from Paligot, the only other comprehensive cross-disciplinary study is the brilliant demonstration by Laurent Mucchielli of the "discovery of the social," centred around Emile Durkheim and his collaborators, and the turn away from the anthropological. While benefiting immeasurably from Mucchielli's studies, I shall show how he may have overestimated this evolution in the case of the psychologist Théodule Ribot and even of some sociologists.[16]

The established university philosophy curriculum was the cradle of the new social sciences of psychology and sociology. Some students of the old introspective psychology and eclectic philosophy curriculum rebelled against pure reasoning wishing to be more empirical or even experimental, and to relate philosophy to biology. Such motivations underlay the career of the founder of French empirical psychology, Théodule Ribot (1839–1916) and also affected some of the authors in the periodical he established in 1876, the *Revue philosophique*.[17]

While Serge Nicolas has discussed psychological heredity and has written a major biography of Ribot, a neglected aspect of Ribot scholarship is the psychologist's interest in the anthropological perspective beyond the individual. Chapter 4 shows how Ribot never quite overcame his fascination for racial hierarchy and cultural evolution, even as he moved in the direction of the sociological. Mucchielli and Bertolini have studied aspects of the *Revue philosophique*, but in chapter 5 I use different evidence to discuss the gradual infiltration of sociological theory

14 Blanckaert, "Le premesse dell'antropologia," 51–70; Hecht, "A Vigilant Anthropology," 221–40.

15 Paligot, *La République raciale*.

16 Mucchielli, *La Découverte du social*.

17 Nicolas, *Théodule Ribot, philosophe Breton;* on the *Revue philosophique* Thirard, "La Fondation de la 'Revue philosophique,'" 401–13; Bertolini, *Il Pensiero e la memoria*; Mucchielli, "Aux origines de la psychologie," 263–89.

and the increased questioning of hard-line hereditarian anthropology. In chapter 6 the study of the famous inventor of the "metric scale of intelligence," Alfred Binet, similarly displays the tantalizing attractiveness of anthropologists' measurements in psychology, though Binet finally turned away from a simplistic overvaluation of anthropology. Elisabeth Chapuis's interpretation of Binet's fascination for the quantitative proves convincing.[18]

The founder of Positivism, Auguste Comte, invented the term sociology and later in the nineteenth century the influence of the systematic philosopher Herbert Spencer helped popularize it all over Europe. Comte's disciple Emile Littré had founded a short-lived Société de sociologie (1872–74) that was more a republican political club than a scholarly research association. However, the institutionalization of sociology in flourishing periodicals occurred after the heyday of Comte and Spencer with the rise of the group around Emile Durkheim beginning in 1896 (*L'Année sociologique*) and also with the more neglected organicist and biologically oriented associates of the *Revue internationale de sociologie* (1893) around René Worms.[19] Through the medium of periodicals and learned societies, I investigate the turf wars and shifting boundaries among these emerging disciplines. Authors such as Snait Gissis label the organicists of the Worms group as believers in racial hierarchy.[20] However, Daniela Barberis has supplied a needed corrective dissociating biological analogics from the necessity of racial theory.[21] A close investigation of the periodicals of the sociologists associated with Worms reveals that many members of the group did not neglect innate elements or choose to abandon entrenched values about gender roles. However, there was a built-in bias among sociologists favouring belief in the influence of the social milieu. Even though they were less flexible on gender than on race, approved social roles were as important as natural endowment in setting limits to the role of women. For most social scientists true egalitarianism on race and gender was still a distant dream at the turn of the twentieth century.

18 Chapuis, "Binet, une passion de la mesure," 71–81; Andrieu, "Céphalométrie scolaire et intelligence inégale," 157–74.

19 On Littré's group, Johan Heilbron, "Sociology and positivism in 19th-century France," 30–62; Logue, *From Philosophy to Sociology;* Mucchielli, *La Découverte du social*; Becquemont et Mucchielli, *Le Cas Spencer*, Terry Clark, "Marginality, Eclecticism, and Innovation," 12–27; Geiger, "René Worms, l'organicisme et l'organisation," 345–60; Mosbah-Natanson, "Internationalisme et tradition nationale," 35–62.

20 Gissis, "Late Nineteenth Century Lamarckism and French Sociology," 69–122.

21 Barberis, "In search of an object," 51–72.

The decade before World War I saw a remarkable change in existing ethnographic and anthropological institutions. The disappearance of an older generation, the increased sensitivity to the excesses of racial theory, and the crisis in measurement standards all helped bring about a move toward studying cultural and linguistic practices in the new Institut ethnographique international and Institut français d'anthropologie, and this trend persisted in a revived Société d'ethnographie. Nevertheless, as the Italian scholar Filippo Zerilli has shown, the physical and the natural were always significant elements even during this institutional shift.[22] Cultural and ethnographic analysis never completely eclipsed the hard-line physical anthropology that persisted in the Société d'anthropologie.

A remarkable revival of racial and gender stereotypes occurred in the 1930s and most notably during the Vichy regime. Anti-Semitism was only a minority belief among nineteenth-century ethnographers and anthropologists, and the case of the anthropologist Paul Rivet shows a strong commitment to combat simplistic racial theory. Still, teachers at the Paris École d'anthropologie attempted to redefine biological criteria for race, such as blood type, and to add an ethnocultural element that would capture biologically diverse groups such as the Jews. These extreme anthropological interpretations and the anti-immigrant hostility of the era precipitated a sinister devaluation of the place of Asians, Africans, and especially Jews in France. Without Vichy such views would never have gained scholarly prestige, but the earlier debates about race and indelible heredity had sown the seeds of a hierarchical view of human groups. Such denigration of "difference" within French society has been a continuing issue whose traces can be seen, albeit without scholarly race and gender theories, in present-day controversies about French identity.

THE CURRENT NATURE-NURTURE DEBATE

The current debate for a non-specialist audience inevitably leads to considerations about how "human nature" (or the lack of a fixed nature) relates to policy decisions. Its most recent incarnation demonstrates the more stereotypical division between conservative hereditarians and progressive environmentalists. The work of Steven Pinker, an expert on genetic brain development and language development, illustrates this aspect of the modern debate. Pinker has employed studies in cognitive science, evolutionary psychology, and behavioural genetics to bolster a "tragic" view of human nature and to invalidate a "utopian" vision stemming from

22 Zerilli, *Il Lato oscuro dell'etnologia.*

Enlightenment optimism and Marxism.[23] He argues that many twentieth-century sociologists and anthropologists erroneously subscribed to a Lockean "blank slate" view of the mind and therefore to the malleability through culture of any human individual. They justly criticized the racial determinism of Nazism but neglected the dangers of assuming the possibility of wholesale social engineering – dangers visible through the consequences of the regimes of Stalin and Mao. There are distant echoes here of the nineteenth-century controversies about the educability or "perfectibility" of humans. Clearly this view would be unfriendly to heirs of the anthropologist Franz Boas and of the sociologist Emile Durkheim who magnified the autonomy of culture and constrained the scope of innate inclinations.[24]

In Pinker's view, the struggle for existence has hard-wired in the human brain inclinations to violence and aggression because of their adaptive importance. Similarly, humans always favour their own genetic kin. Hence, it would be vain to foresee or to attempt to achieve a completely peaceful world or, in view of family bonds, to expect a successful communal lifestyle.[25] In addition, genetically programmed psychological fundamentals limit variation across cultures, as suggested by the ethnographer Donald E. Brown's catalogue of at least 400 allegedly cross-cultural traits. For example, humans generally treat their children well because they're hard-wired to do so, and men are hard-wired to be risk-takers to impress females.[26]

Pinker, even when he explains the success of some psychopathic individuals, does not mean to advocate manipulation, mayhem, or murder. But in line with the tragic view, evolutionary psychology argues that the basis for the evolution of altruism was reciprocity. People help others when they expect personal gain to result. The political implication is that people would resent and resist overly generous welfare programs that increase taxpayers' burdens while providing recipients more than the basic food and shelter necessary for social peace.[27] In the late nineteenth and early twentieth centuries, elements of social solidarity were incorporated even among many hereditarian but socially progressive anthropologists. One of the more conservative disciples of the anthropologist Paul Broca, Paul Topinard, generally avoided the sociological implications of anthropology. Yet near the end of his life he wrote a manifesto that maintained that altruism could be inculcated by hereditary habits, and by direct social guidance to counteract the egotistical passions of "animal man."

23 Pinker, *The Blank Slate*, 281–94.

24 Ibid., 27–84.

25 Ibid., 254–7, 294.

26 Ibid., 37–40, 73, and 435–9 on Donald E. Brown, "human universals."

27 Ibid., 193, 242.

However, somewhat like Pinker, he worried about overly intrusive government actions that would raise taxation and remove the stimulus to individual effort.[28]

The most controversial application of hard-wired traits is sex differences. Pinker argues for male preference for promiscuity and violent competition on the basis of universal evolutionary psychology, and, among a few low-status males, even an inclination to rape based on sexual deprivation. Psychological testing of women leads him to speculate that women are destined to flourish in business careers, while their somewhat lesser talents in spatial relations will make it vain to try to motivate most of them to embark on careers in science and engineering. Hence, as for half-hearted feminists like nineteenth-century anthropologist Léonce Manouvrier, feminism must be moderate or reasonable.[29] The ideological component is that Pinker wishes to call attention to the fact that affirmative action programs always include the possibility of reverse discrimination and elicit resentment of the successful.[30]

In 1875 the British anthropologist, psychologist, and statistician Francis Galton, the originator of the nature-nurture opposition, arrived at the idea of studying twins to determine the relative influences of heredity and environment.[31] Pinker and some modern behavioural geneticists use the study of identical and fraternal twins raised apart or the resemblance of adopted children to biological rather than to adoptive parents to argue the case for hereditary predominance. Personality studies indicate an 80 per cent correlation in identical twins, and apparently indicate that at least 40 per cent of the variation is genetic, and only 10 per cent or less due to family environment, though another 25 per cent is attributable to the unique experiences (circumstances, accidents, etc.) of the individual. The consequence is that, after provision of basic diet, shelter, and normal affection excluding neglect by parents, parenting styles don't seem to matter much. Pinker contends that 50–70 per cent of variations in IQ are heritable.[32]

Disputes continue about the significance of the twin studies, and whether twins reared apart have actually had similar surroundings in

28 Topinard, *Science et Foi*, 3, 51, 348, 397.

29 Pinker, *The Blank Slate*, 337–49, esp. 340–1; for Manouvrier's half-hearted feminism, see Manouvrier, "Conclusions générales," 41–61.

30 Pinker, *The Blank Slate*, 112, 135, 205–7, 353 on brain differences between the sexes.

31 Galton "The history of twins, as a criterion of the relative powers of nature and nurture," *Fraser's Magazine* 12 (1875): 566–76, cited in Burbridge, "Francis Galton on twins," 323.

32 Pinker, *The Blank Slate*, 45–8 on twins, 150, 373, 386.

the early months of life with extended families or in the same village before separation. Also controversial is whether adopted children have higher IQs than their parents for social rather than biological reasons. Raymond Fancher has argued that for twin studies to be valid one needs a seldom-achieved random placement and a guarantee of complete separation at birth. [33]

The most general critique of a modern hereditarian view is that the environment is so influential on the way genes are expressed that one cannot calculate such simple proportions of inheritance versus environment. Geneticist Richard Lewontin argues that the "bucket" metaphor of genetic ability is misguided. The claim is that with equal amounts of water, some genetic "buckets" will simply overflow because they lack capacity. For Lewontin, there is no simple "normal" environment for "tendencies" or "buckets," so that the "constant interaction of the developmental environment" with genes is always more complex. The cultural environment, according to Lewontin, can change by orders of magnitude any measure of mental ability, so that inherited traits are not fixed. Lewontin admittedly is influenced by socialist political convictions, as Pinker has charged, but his arguments are not always invalid because of them any more than Pinker's arguments are invalid because he is more conservative.[34] Lewontin's views have recently won strong support from biologists such as Michael Meaney and in popularized form by the journalist Daniel Shenk. For Meaney, life emerges only from the interaction of nature and nurture, and "there are no genetic factors that can be studied independently of the environment, just as there are no environmental factors that can function independently of the genome."[35]

There is also considerable dispute about whether the aggressive, violent, risk-taking traits useful in the evolution of the human male, for example, are really such hard-wired inclinations, so that the very contents of the brain are pre-programmed, rather than merely the neural networks making such responses possible. Not all traits may be traced back to those that were adaptively selected in the early eras of human existence; they may merely be consequences of brain architecture.[36]

For science journalist Matt Ridley, the most significant conclusion is that genes enable development in certain environments in certain ways rather than predetermine all patterns of behaviour. As Ridley expresses

33 Lewontin, *Biology as Ideology*, 33–5; on the fallacies of Cyril Burt's twin studies, see Fancher, "Science, Ideology, and the Media," 1565–6; see also Shenk, *The Genius in All of Us*, 59–68.

34 Lewontin, *Triple Helix*, 28–30, 35; Lewontin, *Biology as Ideology*, 26–9.

35 Meaney, "Nature, Nurture," 51; Shenk, *The Genius in All of Us*, esp. 14, 19, 27.

36. Gould, "Evolutionary Psychology: An Exchange," 55–6.

it, there was a co-evolution of brain and culture, in which traits such as intelligence, empathy, foresight, co-operation, betrayal, and deceit developed before art, adornment, crafts, and trade.[37] Neurobiologist Joan Stiles contends that the inherited and environmental factors in brain development are inseparable. The developmental process depends on the interaction of inherited and contextual factors – neither nature nor nurture has exclusive jurisdiction. Environmental factors, including modifications by diet and stresses such as alcohol, drugs, or radiation may lead to pathology of brain development.[38] For Stiles, as for many nineteenth-century thinkers, the innate and environmental clearly act in concert and separating them is a vain exercise. However, unlike many earlier thinkers, she does not exert substantial effort to assert the superiority of either nature or nurture. Moreover, in recent years research on the "markers" for expressing genes ("epigenetic" matter) has shown that behavioural practices such as smoking may influence the longevity or proneness to cardiovascular disease of future generations, and therefore the "epigenetic" expression of genes, even without direct mutation of the DNA. In this sense, there are more direct influences of environment on phenotypic outcomes than previously believed.[39]

Pinker himself also agrees with the pathbreaking study of Jared Diamond concerning the reasons for differential cultural achievement (or at least expansionism) in societies relating to the physical environment of cultivable plants and domesticable animals, and other geographic factors.[40] Yet his general argument assesses these limits as less important than the genetic predispositions independent of the environment.

The core issue of whether there is an identifiable human nature remains, with the corollary of how to react to its reputed features. If certain adaptive qualities are hard-wired across cultures, how should one promote or repress them? As in race and gender issues in the nineteenth and early twentieth centuries, ideology will often intrude upon the expected answers.

For example, Steven Pinker, despite his fervent wish to restrain psychopaths and rapists and to punish criminals, ultimately argues that aggression is a cross-cultural trait that cannot be eradicated. In a hereditarian view, realism about aggression militates against overly high expectations for a peaceful world. Similarly, he strongly condemns racism and discrimination against women, but he believes there are statistical differences in average gender abilities in such skills as perception

37 Ridley, *Nature via Nurture*, 225, 230, 250.

38 Stiles, *The Fundamentals of Brain Development*, 13, 15, 62, 365.

39 Whitelaw, "Epigenetics," 131–2.

40 Pinker, *The Blank State*, 67–9; see Diamond, *Guns, Germs, and Steel*.

in spatial relations.[41] If Pinker expects answers for policy decisions given by a concept of human nature, what guidelines would he propose for promoting incipient talents if aspiring young women wish to be engineers? If there were no weighted guidelines and no predictive power for individual cases, how would the alleged statistical difference affect decision-making?[42]

If such debates are still prevalent, the arguments in previous centuries about nature and nurture are likely to continue. In France of the Third Republic the disputes assumed critical dimensions because they affected fears of depopulation and gender roles, immigration and the integrity of French nationality, alcoholism and mental disturbance, and the relationship between France and the rest of the world. The expansion of French control in Africa and southeast Asia kept, at least for the political elites, issues of empire very much a vital concern, even if the French population at large was often indifferent. Moreover, the perpetuation of stereotyping of non-Europeans in school manuals inculcated existing cultural prejudices in the population, often with a patina of "science."[43] Increasing clamour from the 1880s for Civil Code reform, and for greater civil and political rights for women propelled gender issues to the forefront of public attention.

Here I do not investigate all aspects of the heredity-milieu debate in the formation of the social sciences. Existing literature on criminology and on the Durkheimian sociologists has already substantially developed these subjects. Nor do I explore early incarnations of eugenic thought, but rather show how the controversy permeated discussions on race and gender in the emerging social sciences. The anthropological perspective that so influenced the other social sciences itself entwined heredity and milieu in the widespread belief that the milieu elicited changes that were then inherited. The founder of empirical psychology Ribot appropriated this anthropological concern before fitfully and half-heartedly surrendering it. Sociology showed a natural bias toward considering the social milieu as important, but the "organicist" sociologists were reluctant to give up biological analogies.

41 Pinker, *The Blank Slate,* 124, 177–9; in more recent work, however, without surrendering his confidence in an identifiable and relatively intractable "human nature," Pinker is developing a theory of the influence of political and social institutions on the decline of personal violence; Pinker, "A History of Violence," 1–6.

42 For a recent debate, see Steven Pinker and Elizabeth Spelke at Harvard in the "Science of Gender and Science."

43 Aldrich, *Greater France*; Schneider, *An Empire for the Masses;* Wilder, *The French Imperial Nation-State*; Paligot, *La République raciale* on Jean-Marie-Antoine de Lanessan, Paul Bert, and school manuals, 70, 138–41.

Moreover, the most hereditarian did not always advocate the most uncompromising imperial expansion, while partisans of the milieu could be apostles of empire. Yet the latter group did not necessarily embrace an egalitarian framework on questions of race. Nor did advocacy of the educability of European women assure gender equality. This study also questions the inevitability of the equivalence of all the marginalized. The general argument has been that advocates of inequality between the sexes used analogous arguments to those used by supporters of racial hierarchy. In the debates among Paris sociologists, however, the race and gender issues did not always elicit consistently conservative or progressive viewpoints.[44]

Among the first to recommend the scientific study of human cultures was the Société d'ethnographie de Paris, whose guiding spirit was an expert on Asian languages and cultures, Léon de Rosny. Their discussions revealed the difficulty of being independent of cultural prejudices about nature, despite their overwhelming concern with the culture of established or vanished civilizations.

44 Stepan, "Race and Gender," 261–77.

2

The Ethnographers

THE ORIGINS OF ETHNOGRAPHY

THE STUDY OF THE HISTORY and culture of peoples long predated the foundation of scholarly societies devoted to it. In the eighteenth century such scholarly study was the province of philosophers and travellers' reports. With the advent of Carl von Linné (Linnaeus) in Sweden and Georges-Louis Leclerc, comte de Buffon in France, naturalists studied human varieties as products of different environments.[1] The terms "ethnology" and "ethnography" originated in the German-speaking world. In 1771–72 August Ludwig Schlözer of the University of Göttingen used "ethnography" as a synonym for Völkerkunde (all peoples) or Volkskunde (one people) in a universal history of peoples of the globe, including accounts of travellers and comparative linguistics. Eleven years later the historian of public law Adam Kollar employed "Ethnologie" or "Ethnologia" in a Latin work to denote a descriptive, historical study of the origins, language, and customs of peoples. In 1826 the Italian linguist Adrien Balbi published in France an *Atlas ethnographique du globe* equating ethnography with a classification of all peoples in language families.[2] Before mid-century, "ethnologists" and "ethnographers" would rely less on language and culture and more on differentiation of physical types.

While members of geographical societies might undertake ethnological study in the early nineteenth century, during the monarchy of Louis-Philippe a specialized society emerged for exploring the nature and culture of human beings. In the decade after 1839, in response to British philanthropic concern for aboriginals, and increased interest in racial

1 Blanckaert, "La valeur de l'homme," 583–600.

2 Vermeulen, "Origins and institutionalization of ethnography," 50–5; Balbi, *Introduction à l'Atlas ethnographique*, xvii, lxv, lxxxi.

theory, Jamaican-born William Frédéric Edwards established the Société ethnologique de Paris. Edwards had expressed his own interest in observing "forms and proportions of the head and traits of the face" of various French and European regional populations.[3] He reacted favourably to the historical interpretations of Augustin and Amédée Thierry that pitted the descendants of the ancient nobility against the descendants of the bourgeoisie, and tended to view the progressive and democratic factions as the triumph of the Celts over the Franks.[4]

The mission statement of the Société ethnologique de Paris proposed to study "physical organization, intellectual and moral character, languages and historical traditions" of the "human races." They would further establish through collections of portraits and objects the "degree of intelligence and culture that distinguishes diverse peoples."[5] Edwards and his colleagues were more concerned with races and prefigured physical anthropology, but they also wished to investigate material culture and customs. For Edwards human types were permanent, with changes coming almost entirely from crossbreeding. Interbreeding races would produce intermediate forms only when the two original types were quite distant from each other.

In their extensive discussion on the capabilities of "whites" and "blacks" in 1847, the society revealed a strong organicist Saint-Simonian current that seemed to overcome its weaker egalitarian voices. In this view, inherent physical qualities of Africans would make it less likely for them to progress without the guidance of Europeans. Opinions were not so extreme as to consider Africans subhuman or incapable of any progress, but the majority thought it not very likely that Africans could achieve the degree of civilization now commonplace in Europe.[6] The society clearly believed in the importance of a civilizing mission as well as the indispensable guidance of Europeans even before France undertook substantial imperial adventures on the African continent or anywhere outside the south Pacific. The political distractions of the revolutionary era of 1848 left this society in a moribund state. Anthropological and ethnographical questions in the subsequent decade reverted to the

3 Edwards, *Des Caractères physiologiques des races humaines*; Edwards, "Esquisse de l'état actuel de l'anthropologie," 124–5; Blanckaert, "On the Origins of French Ethnology," 34–45, esp. 44–5.

4 Rignol, "Anthropologie et progrès," 157–79; Rignol et Régnier, "Races et politiques dans l'Histoire de France," 127–52.

5 *Mémoires de la Société ethnologique de Paris* 1, i, iii, iv; "Instruction générale adressée aux voyageurs," vi-xv.

6 Staum, *Labeling People*, 122–57.

Société de géographie de Paris or the venerable Société asiatique, and to a brand-new Société des américanistes.

TWO RIVAL SOCIETIES

The conservative and repressive intellectual atmosphere of the Empire of Napoleon III did not stifle the foundation of two new scholarly societies, ostensibly quite different from each other, though with perhaps more covert resemblances than most scholars have previously acknowledged. The founders of the new Société d'ethnographie américaine et orientale (later more commonly known as the Société d'ethnographie) hoped to help regenerate Europe with the wisdom and religious revelations of the manuscripts and religious classics of a broadly defined Orient (including north Africa, west Asia, and Turkish Europe), though they seemed to have little doubt about ultimate European superiority. They also wanted to study the ancient cultures of pre-Columbian America to supplement the work of the Société des américanistes.[7]

The foundation of the second scholarly society is a better-known story. In the tense atmosphere of concern about offence to the Church, the Société de biologie found a reading by the illustrious physician and anatomist Paul Broca of a memoir on hybridity too sensitive. Broca's goal was to show, with examples from the animal kingdom, that there was no simple answer to the question of whether the fertile offspring of different human "races" proved the unity of the species.[8] Stymied by the biologists, he therefore convinced a group of physician and naturalist friends to found their own Société d'anthropologie de Paris, which held its first official meeting in May 1859. Broca had been a committed republican, although disillusioned with the turbulence of the populace in the revolution of 1848.[9] Still the anti-slavery debates among the earlier Société ethnologique and the possibly secularist and anti-Biblical overtones of anthropological study assured a presence of police spies for the first few years at meetings of the society. Among other reasons, the authorities mistrusted members of the medical faculty because the hotheaded

7 Rosny, "L'Orient," 48–59, esp. 59; Chailleu, "Histoire de la société d'ethnographie," 89–107, esp. 95–7.

8 Blanckaert, "Of Monstrous Métis?," 42–70.

9 On Broca, see Schiller, *Paul Broca;* Blanckaert, "Préface," i-xliii; on the anthropologists, Williams, "The Science of Man"; See Williams, "Anthropological Institutions," 331–48, on the failure to institutionalize ethnography, 343–4; Harvey, "Races Specified, Evolution Transformed"; Hecht, *The End of the Soul;* Blanckaert, ed. *Les Politiques de l'anthropologie;* Wartelle, "La Société d'Anthropologie de Paris," 125–71; Paligot, *La République raciale.*

younger staff had rebelled against the dean.[10] The preponderance of physicians and naturalists in this society and the long tradition of relating physical and mental aspects in medicine would assure the dominance of physical anthropology. But once again programmatic statements also asserted interest in culture, linguistics, and archaeology.

The organizing spirit behind the ethnographers was the twenty-two-year-old linguist Léon de Rosny (1837–1914), son of an aristocratic but pro-republican linguist and student of Latin America. Once librarian of the Société asiatique and contributor to their journal, the younger Rosny studied Oriental languages at the Ecole spéciale des langues orientales, where he would teach Japanese from 1863 and became a professor in 1868.[11] At the Collège de France he studied Egyptian archaeology and Chinese language. He assisted the merger of Orientalists (Amis de l'Orient) and Americanists (first Société des américanistes), who would eventually adopt a seal of three figures from Asian, African, and European "races" with the caption "diverse in body, but brothers in the mind."[12] From 1886 Rosny also taught Far Eastern religions at the Ecole pratique des hautes études and became a protagonist of an "eclectic Buddhism." Rosny was the indispensable society officer – secretary of the group from 1859 to 1872 (except for a hiatus as interpreter to Japanese visitors in 1862) and again from 1880 to 1885, and president in 1873–80, 1889, 1891–92, and 1896–98.

The society patrons and founders included a few wealthy philanthropists as well as holdovers from the earlier Société ethnologique. Among the patrons was a Russian émigré and philologist, Prince Michael Vlangali-Handjéri, who became the second president of the society and first president of the council. The first president was a former diplomat, Baron Paul-Charles-Amiable Bourgoing (1791–1864) who ingratiated the society with the Imperial government through his survey of the region of the new French protectorate of Cochinchina. From the earlier ethnologists came the Institut de France librarian and later member of the Academy of Inscriptions, Louis-Ferdinand-Alfred Maury (1817–1892), an avid student of myths, religions, and a theorist of race and language. Two former members of the ethnologists were reasonably active in the new society: the Egypt expedition veteran and expert on

10 Blanckaert, "La Crise de l'anthropométrie," 101.

11 Archives nationales (hereafter AN) 62 AJ 5, Assemblée des professeurs; Rosny had only one student in 1895.

12 Comprehensive summary in Williams, "The Science of Man," 130–80; Lacombe, "Essai sur les origines et les premiers développements de la Société d'ethnographie," 329–41; Chailleu, "Histoire de la société,"; on Rosny, see also Chailleu, "Léon de Rosny," 85–107; for increasing nationalism in the society, see Edison, "The Society of Ethnography in the Nineteenth Century."

African geography, Edme-François Jomard (1777–1862), president at the time of his death; and the Hindi linguist and devout Catholic, Joseph-Héliodore-Sagesse-Vertu Garcin de Tassy (1794–1873), who was president in 1864–65. Two noted Americanists of the early years typified the work of the society. The Belgian-born chaplain of the French legation in Mexico, abbé Charles-Etienne Brasseur de Bourbourg (1814–1874), and his colleague at the Sorbonne Faculty of Letters Joseph-Marius-Alexis Aubin (1802–1891) both deciphered pre-Columbian Mexican glyphs and wrote histories of early peoples of Mexico.

Jomard died too soon to influence the fate of the society, but he is a special case in the history of ethnography. Ever since 1831 he had proposed a plan for an ethnology museum that exhibited thematic categories of the implements, crafts, and clothing of various peoples from around the globe on their way to civilization, along with their physical likenesses and languages.[13] In a tortuous bureaucratic pathway, Jomard's early plans helped the establishment of what became the Musée d'ethnographie at the Trocadéro in 1878. The artifacts would "form a series of materials suitable to give a correct idea of the different human races classified according to their physical characteristics as well as the moral and intellectual notions and the fruits of their industry."[14] For Jomard in 1839 "ethnology is the science of geography itself seen as a whole and in its highest generality." It was "urgent to know thoroughly the degree of civilization of all races," both in physical constitution and cultures, to "have relations between them and us which are more certain and advantageous."[15] Jomard was always interested in both nature and culture – physical measurements and artifacts.

At least thirteen of the early membership were either Institut de France members, or from the Collège de France or Ecole des langues orientales vivantes. One of the most famous scholars on the council was the Semitic languages and religious studies scholar Ernest Renan, who had pointedly refused an offer to be an active participant in the Société d'anthropologie.[16] Although Renan was also inactive with the ethnographers, he represented the ideal of the superiority of cultural and linguistic methods to those of physical anthropology. Still, he had strong views about the relative capabilities of superior "Aryan" and less

13 Dias, *Le Musée d'ethnographie du Trocadéro (1878–1908)*, 15, 115–32; Dias, "Une Science nouvelle," 159–83, esp. 161, 165, 167–8, 170–1, and 177 on "ethno-geography."

14 [Jomard], *Considérations sur l'objet et les avantages*, 19–20.

15 Jomard, *Etudes géographiques et historiques sur l'Arabie* (1839), cited in Blanckaert, "'Story' et 'History' de l'ethnologie," 461.

16 Archives de la Société d'anthropologie de Paris, Musée de l'homme, Registre d'inventaire, 1B1; Blanckaert, "Les deux parties du problème," 62–89.

gifted "Semitic" races and about the hereditary moral and intellectual character of a people, particularly in "primitive" times before the influence of religion and civilization.[17]

One of the most eminent council members, who was president in 1862–63, was the physiologist Claude Bernard, who praised Rosny's methods. Bernard's talk in August 1862 advocated both physical and cultural approaches to the study of ethnography amid a triumphalist exhortation to "bring to peoples in childhood the benefits of civilization," teach them moral laws, and through a "community of exchanges in physical and moral order make men better and happier."[18] Later he made clear that his physiology was not to be considered materialist, as it was only concerned with "organic and intellectual conditions for thoughts and feelings."[19] However, Bernard's association with the society was too fleeting to add to their prestige in the general scholarly community.

From the political and business worlds came diplomat and entrepreneur Baron Jules de Lesseps, an administrator in north Africa, who was president in 1866. The politically orthodox and socially elitist nature of the society in its early years seems illustrated by its eagerness to support French colonial activities in Cochinchina. Moderate Republicans, like Rosny himself, however, were never out of place. An enthusiastic supporter was the ex-Saint-Simonian, education minister of the Second Republic, and then senator, Hippolyte Carnot (1801–1888), president from 1880 until shortly before his death. Even near the end of the society's activity, the deputy, government minister, prime minister, and patron of "solidarism" (a non-socialist philosophy of social solidarity) Léon Bourgeois assumed the presidency in 1900. The openness of the society to foreign members was obvious as they collected an assortment of Europeans as well as representatives from Persia and Japan. Rosny successfully gathered delegates for a first International Congress of Orientalists in 1873 and an International Congress of Americanists the following year. Moreover, the society took the initiative in co-ordinating international congresses of ethnography in conjunction with the Paris international expositions of 1878, 1889, and 1900.

The religious bent of the new society, especially as contrasted to the more secular anthropologists, is apparent in the contributions of editors and correspondents from the religious press, such as Biblical scholar and student of religions of India Charles Schoebel (1813–1888), who wrote for the ultra-Catholic *L'Univers* of Louis Veuillot. Augustin de

17 Desmet, "Abel Hovelacque et l'ecole de linguistique naturaliste," 64; Paligot, *La République raciale*, 160–1.

18 Claude Bernard, "Coup d' œil sur la science ethnographique," 285–6.

19 Claude Bernard, "Exposé des faits," 249.

Bonnetty (1798–1879), founding editor of *Annales de philosophie chrétienne* and a student of Tibetan Buddhism and ancient Roman religion, was a believer in the Revelation latent in all religious myths and fables. As late as the 1890s, when the scientific materialists had triumphed in the Société d'anthropologie, the ethnographers reaffirmed faith in deism or in some cases in Christianity as a superior religion.[20]

How then could historians and even contemporaries portray a society so seemingly comfortable with the political establishment and with a substantial complement of members as marginalized?[21] As Elizabeth Williams long ago argued, there was never great theoretical coherence in the papers read to their society, as their difficulties in hammering out a mission statement testified. Nor did they ever establish a school, a museum, or the equivalent of a laboratory. At the time of the 1889 International Congress, the society offered proposals for courses for an "ethnographic school" – general ethnography, ethnology, ethics or Ethnodicée, ethnographic psychology, ethnographic pathology, Buddhism, comparative religion, and Mexican language and archaeology – but this plan remained stillborn.[22] Furthermore, neither the Collège de France nor the Ecole des Langues orientales looked favourably on the establishment of a course of ethnography. The Ecole pratique des hautes études enabled some religious studies teachers to include ethnographic data, but with no degree of consistency.[23] The ethnographers were also perennially short of funding, ran their publications on a shoestring with considerable delays, constantly reorganized their committees and publications, and did practically no field work on their own.

The ethnographers' situation made for invidious comparisons with an anthropology society with its own private lab housed in abandoned facilities of the Paris Faculty of Medicine in 1867 (after 1868 in the Ecole pratique des hautes études) and its own museum of craniological and other specimens based on Broca's own skull collection after 1872. Most importantly, in 1876 the anthropologists received 12,000 F from the city of Paris and in 1878 20,000 F from the national government to open their own private teaching establishment with six professors, the Ecole d'anthropologie.[24] At first the Ecole taught general anthropology, physiological anthropology, ethnology, linguistic anthropology, prehistoric

20 Desdouits, "Le Progrès moral et ses conditions," 146–63; Réthoré, "Plaidoyer pour le dogmatisme," 159–61.

21 Stocking, "Qu'est-ce qui est en jeu dans un nom?" 421–32, esp. 423.

22 *Revue orientale et américaine* 2nd series, 13 (1889): 137; Williams, "The Science of Man," 160–70 and "Anthropological Institutions," 338.

23 Ibid., 345.

24 Williams, "The Science of Man," 160–6; "Anthropological Institutions," 338.

anthropology, and demography and medical geography. Over 8,000 auditors attended nearly 200 lectures in 1877, nearly 11,000 in 1897–98, and an average of between 10,000 and 13,000 from 1893 to 1907.[25] However, by 1911 the professors were lamenting their diminishing enrolment and financial difficulties.[26] The Société d'anthropologie had 659 members and associates in 1879; in 1885 it reached a maximum of 757 members (459 titular), but declined to 405 members (235 titular) by 1914. The ethnographers registered 115 members in 1864 and 132 (34 titular, 27 "free" members, and 71 correspondents) in 1878 (though some scholars claim 470). In 1889 the society had 450 titular members and 149 associates, and, in 1890, 871 voted for president (375 from France). But there was a sharp decline in 1893 to 184 regular members, and by 1903 only 165 regular members remained.[27]

Besides institutional weakness, Rosny's eccentric notion of scientific respectability in an era of trust in quantitative measurement included his intuitive "méthode conscientielle," an internal feeling concurring with the observations of the senses.[28] Moreover, his quarrelsome nature assured failure in obtaining either the Chinese language chair at the Collège de France in 1893 or election to the Academy of Inscriptions in 1897.[29]

The most important theoreticians in the society, aside from Rosny himself, Jean-François-Charles dit de Labarthe (1812–1871) and Maurice-Alphonse Castaing (1822–1888), were also eccentrics relative to the French scholarly community. Both illustrate the reputation of the society as valuing culture over craniology. Yet both paradoxically were committed to a physically based introduction to ethnography. Labarthe was a mathematics teacher and member of the Société asiatique who learned Chinese at the Collège de France and introduced Rosny to the Chinese classical religious texts. Later he would develop expertise in the comparative grammar of Sanskrit and Zend, pre-Columbian Inca and Mexican culture, archaeology, and comparative religion. Like the Rosny family, he had been an enthusiastic supporter of the Republic in 1848, but he accepted revelation in all religions.[30] Castaing, a lawyer and War Ministry attaché with an interest in civil engineering, provided some of

25 Paligot, *La République raciale*, 124, 130.

26 Williams, "Anthropological Institutions," 336, 344; Bernardini, *Le darwinisme social (1859–1918) en France*, 149–50; Zerilli, *Il Lato oscuro dell'etnologia*, 46.

27 Sibeud, *Une Science impériale pour l'Afrique*, 38, 50; Wartelle, "La Société d'anthropologie," 132–3, 153–5.

28 Williams, "Science of Man," 178.

29 Luc Chailleu, "Léon de Rosny," 103–4.

30 Rosny, Obituary of Labarthe in *Actes de la Société d'ethnographie* (hereafter *Actes*) 8 (1874–77): 369–75.

the most anti-anthropological and anti-racist statements on record, while at the same time he seemed covertly to disavow them. Dabbling in the ethnography of his native Aquitaine and later writing a general French ethnography for the schools, he expressed interest in Native American mythology and the Native American place in the civilized world, the religion of the Incas, and, as an Orientalist, in Morocco and the Druze of Syria and Lebanon.

The conventional and largely justifiable contrast of the two societies rests on their differing notions of proper scientific method, their divergent political and religious sensibilities, and their unhappy relationship. The ethnographers were hostile to craniology, studied only texts, and did little fieldwork. Many members respected traditional religious outlooks and abhorred materialism. By contrast, the anthropologists privileged measurement of crania or facial features, issued instructions to travellers, and corresponded with a large circle of military and naval physicians to gather data. After Broca's death, contrary to the wishes of some of his heirs, a group of combative scientific materialists dominated the society and advanced a secularist, anti-clerical, and evolutionist program intended to aid social reform within France and make the enterprise of colonization scientific.

The largely conservative ethnographers hoped to build an anti-materialist "science of destiny." Rosny invented the term "ambiantisme" for a philosophy that ostensibly privileged the cultural milieu rather than the inborn nature of peoples.[31] The programmatic statements of many ethnographers expressed fraternity and confidence in the virtual equality of all peoples.

There is little doubt that the two societies never co-operated. Rosny unsuccessfully implored his friend, the ethnographically inclined anthropologist Ernest-Théodore Hamy, assistant naturalist at the Muséum, to facilitate mutual exchange of society publications. Hamy finally does appear on a membership list of the society in 1890, though he was never active.[32] The anthropologist physician Georges Hervé refused an exchange of publications with the African Society because he distrusted the "personality of the editor," Eugène Verrier, a prominent Société d'ethnographie physician who had resigned in 1890 from the Société d'anthropologie under a cloud of criticism for inaccuracy of maps.[33]

31 Rosny, *Traité d'ethnographie théorique et descriptive*, 1: 54–5.

32 Correspondance de Hamy, Muséum national d'histoire naturelle MS. 2254, Lettre de 16 février 1873.

33 Archives de la Société d'anthropologie, Procès-verbaux du comité central, 1890–1901, 12 mars 1891.

Moreover, the anthropologists often devalued the publications of the ethnographers. Although Hamy approved of Rosny's studies of "the Siamese" in 1873, two years later Cyprien Issaurat rather minimized the importance of the ethnographers' memoirs for his fellow anthropologists.[34] Yet an exploration of the periodical publications of the societies and the very limited usable archives suggests a re-interpretation of the differences between the two groups.[35]

JURISDICTIONAL CONFLICT

While both ethnographers and anthropologists saw some role for heredity and environment, each society refused to surrender the study of the area they conceived as secondary. When Rosny in 1878 recalled Castaing's proposal for a program for the ethnographers back in 1860, he repeated the mantra that ethnography is the "physical, moral, and intellectual study of man."[36] The anthropologists, especially those associated with the Muséum national d'histoire naturelle, also refused to surrender the cultural aspects. They deployed two different strategies – either to define the ethnographers' works as merely descriptive, and therefore not really scientific, or to classify them as a completely auxiliary study to the main work of anthropology.

The ethnographers also never gave up the desire to conduct physical studies even though they rarely implemented their plans. Both Castaing and Labarthe charted a program for ethnography that included physical anthropology as a necessary introduction to the more important historical study of peoples. Labarthe's 1862 program had a lengthy "ethnographic" classification of races. Labarthe thought ethnography superior to anthropology because it was not marred by materialist determinism. It also dealt with the study of the milieu, or "mésologie," (a term invented by the demographer and member of the Société d'anthropologie Louis-Adolphe Bertillon) both in the sense of physical regimen and culture – language, customs, religion, and social institutions.[37] Ethnography would be a "science of destiny" that would help eradicate racial prejudice and

34 Hamy, "Revue des journaux," *Revue d'anthropologie* 2 (1873): 137–44; Issaurat, "Sur les Mémoires de la Société d'ethnographie," 192, cited also in Williams, "Anthropological Institutions," 338.

35 The archives of the nineteenth-century Société d'ethnographie have disappeared and the archives of the Société d'anthropologie are in poor condition so that even the archivists can no longer find some documents used by Elizabeth Williams and Jennifer Hecht.

36 *Congrès international des sciences ethnographiques*, 1878, 36–7; "Programme de 1860," *Actes* 8 (1874–77): 70–92.

37 Labarthe, "Aperçu general de la science ethnologique," 342, 346, on anti-materialism; "Essai sur les classements bibliographiques," 8–9; Cf. the huge ambitions of Labarthe and

historical injustice. This ambition induced Rosny to warn against introducing too much anthropology as a distraction.[38]

At the International Congress of Ethnographic Sciences in 1878, the irrepressible, non-conformist Darwinist member of both societies, Clémence Royer, irked her colleagues by claiming "ethnography is a branch of anthropology," its "moral and intellectual side." Rosny countered that anthropology, far from having greater scientific claims, had no place for the moral sciences. On the contrary, anthropology was a branch of ethnography, which at least reserved a place for individual liberty and the dignity of the human species. Castaing similarly argued that ethnography encompasses all of intellectual man as well as having a basis in the physical.[39]

A solution of jurisdictional conflict could have occurred with the exclusion of the so-called "uncivilized" from ethnography. The Sanskrit and Assyrian linguist at the 1889 Ethnographic Congress, Jules Oppert, defined ethnography as the "intellectual movement of human groups constituted in society."[40] This emphasis on the progress of organized groups would have relinquished or minimized the imperative to investigate "savages." However, this definition was proposed in the same era that both Castaing and Verrier were re-introducing anthropological components to the ethnographers' discussions. Meanwhile, anthropologists such as Armand de Quatrefages had never surrendered the cultural domain, while anthropologists Charles Letourneau and Léonce Manouvrier were emphasizing the importance of sociology to the field of anthropology.

The founder of the Société d'anthropologie Paul Broca represented the strategy of defining ethnography as less than a science. In 1864 he called ethnography a clearly secondary "description of peoples," while ethnology, which deserved to be a branch of anthropology, was a "science of human races."[41] Two years later, he added that ethnology included the study of physical resemblances and differences of human types, as well as a comparison of their social states and perfectibility, or

---

the scientific materialist anthropologists in Hammond, "Anthropology as a Weapon of Social Combat," 118–32.

38 Labarthe, "Esquisse d'un tableau préparatoire génésiaque," 146–57; Rosny's comments, 159–61.

39 *Congrès international des science ethnographiques, 15–17 juillet 1878*, 1–8; Harvey, *Almost a Man of Genius*; *Congrès international… 1878*, for Royer, 438, Castaing and Rosny, 440–2, 445.

40 *Congrès international des sciences ethnographiques, 1889*, 8 (Oppert), 26–7 (Rosny); cf. Jomard, "Classification méthodique," 131–7, 241–53; Dias, *Le Musée d'ethnographie*, 23–41 for definitions of ethnography.

41 Broca, "Histoire des travaux de la Société d'anthropologie (1859–1863)." In *Mémoires d'anthropologie*, 415 (vol. 2 in original).

the ability to conceive or receive progress.[42] Broca never excluded from anthropology proper the role of the milieu in modifying the individual or race – that is, of climate, food, and physical and intellectual education. Equally significant was the investigation of "monuments, history, tradition, and religions." He was concerned about the limits within which "the laws of heredity can maintain these variations."[43]

However, he continued to denigrate ethnography by calling it either the "elements of studies rather than a proper science" or alternatively a "historical science" of artificially constituted groups and peoples, whose natural mixture had to be deciphered by trained anthropologists. Only in the direct observation of peoples would cultural information about language, history, and religion be relevant.[44] Broca's most enthusiastic disciple, Paul Topinard (1830–1911), reinforced this boundary in warning ethnographers that they should not try to define races, since "this characterization rests essentially on anatomical bases and is the exclusive domain of anthropology proper, or the zoology of man."[45] Milieu might play a role in crossed races, since mulattoes were darker in the tropics.[46] However, he asserted that the ethnography of customs, religions, and languages, is often "done in a fantastic manner." He chastised his colleague Julien Girard de Rialle, the author of a linguistic study of African and American peoples and a member of both societies, for slighting the fundamental analysis of physical characteristics.[47] It was the "anatomy of brain structure and consideration of the moral and intellectual productions of organs and faculties" that would furnish underlying proof for ethnography that was the mere observation of racially undifferentiated peoples.[48] Hence, the truly trustworthy information came from anatomy. As late as 1907 the ethnography lecturer at the Ecole d'anthropologie, Moindron-Sigismond Zaborowski put forward the same view that "ethnography is not a science independent from anthropology" since craniometry could solve the problem of the origins of peoples. For Zaborowski,

42 Broca, "Anthropologie," *Dictionnaire encyclopédique des sciences médicales*. ed. A. Dechambre (Paris: Asselin, 1866). In Broca, *Mémoires*, 1: 9–10.

43 Broca, "Histoire des travaux," *Mémoires*, 415.

44 Broca, "Discussion sur l'ethnologie et l'ethnographie," 219–20, 223; in 1877 the strict hereditarian Eugène Dally at the Ecole d'anthropologie wished to substitute "demography" for "ethnology," but was more open than Topinard to sources from languages, traditions, and religions; see "Ethnologie" in *Revue géographique internationale*, summary by Ludovic Martinet, 2 (1877): 27–8, 61–2, 68.

45 Topinard, "Anthropologie, ethnologie et ethnographie," 215.

46 Topinard, "De la Notion de race en anthropologie," 654, 658.

47 Topinard, Review of Girard de Rialle, "Les Peuples d'Afrique et d'Amérique," 673–6.

48 Topinard, *Eléments d'anthropologie générale*, 162, 168–9, 175–6.

ethnographers made the mistake of distinguishing peoples by languages or customs alone, whereas they needed the data of racial distinctions that constituted a true ethnology. Ethnography was only a division of anthropology that "ends up determining their place and rank *genealogically and hierarchically*." But a complete science of man, he allowed, would include the acquisition of elements of civilization, such as housing, weapons, ornaments, and clothing. In the same year, Georges Hervé, the ethnology lecturer at the Ecole d'anthropologie, reiterated that peoples of inferior morphology had inferior societies. In an attempted refutation of those who criticized the concept of race, he argued that organic and intellectual characteristics correlated with external physical characteristics. It would be equally false to postulate racial parity as it was to assert the irremediable infirmity of a race. Therefore, he wanted neither the "ethnic imperialism of the Gobinian school" nor the "anarchy of systematic deniers of race."[49]

Some anthropologists, however, refused to remain content with occupying the domain of nature while conceding history to ethnographers. For monogenists intent on showing the migration and adaptation of human races, the evidence of language and ethnography could be important. In 1883 Quatrefages expressed the opinion that anthropology should investigate "intellectual, social, moral, and religious characteristics" of populations.[50] In his 1877 lecture on ethnology, the "science of races," the hard line hereditarian Eugène Dally had also suggested that anthropology would use traditions and artifacts, the data of "ethnography, the science of peoples," not just physical measurements, as its sources.[51] Charles Letourneau, who taught sociology and the history of civilizations at the Ecole d'anthropologie, had prepared a questionnaire on "sociology and ethnography" for the Société d'anthropologie in 1883.[52] The anthropologist most skeptical of excessive reliance on craniometry, Léonce Manouvrier, complicated the issue by insisting no such division of nature from history was feasible. In the periodical of the scientific materialists, *L'Homme*, Manouvrier attacked Paul Topinard for denying that it was essential to know peoples rather than races. While also questioning Dally's suggestion to replace ethnology as the science of races with sociology, Manouvrier asserted that all anthropological sciences have linguistic and sociological aspects. Hence, ethnographers

49 Zaborowski, "Ethnographie," 65; Hervé, "Ethnologie," 77–8; Zerilli, *Il Lato oscuro*, 98–9.

50 Quatrefages, *Hommes fossiles et hommes sauvages. Etudes d'anthropologie* (1883), Paris: Jean-Michel Place, 1988 reprint, viii, cited in Zerilli, "Il Dibattito sul meticciato," 237–8.

51 Dally, "Histoire de l'ethnologie," summary by Ludovic Martinet, 27–9, 61–2.

52 Blanckaert, "Le Premesse dell'antropologia 'culturale' in Francia," 51–70.

needed to learn about races, and anthropologists about peoples. Therefore, it was fruitless to distinguish excessively natural history and history proper. By 1894 Manouvrier was convinced that anthropology was an indispensable guide to legislators, moralists, and educators in its psychological and sociological components and that Broca himself, in comments on the importance of marriage as an institution, had recognized the scope of anthropology.[53] In 1904 he argued that while anthropology studied biological, psychological, and social characteristics, sociology studied social phenomena in both men and animals. The two domains interpenetrated, though they remained logically distinct.[54]

This eclectic view also permeated the new, non-theoretical *Revue d'ethnographie* founded by Hamy in 1882, which wished to expand the boundaries of anthropology to include history as well as prehistoric archaeology, but found the ethnographers and historians reluctant to use craniology.[55] Neither anthropologists nor ethnographers wished to surrender jurisdiction over both physical and cultural aspects of humanity. As the sociological approach gained attention in the era of René Worms and Emile Durkheim, the countercurrent to the anthropological mainstream represented by Manouvrier recognized the importance of the cultural while refusing to give up head measurement.

PHYSICAL ETHNOGRAPHY AND RACE

The conventional picture of the ethnographers stems from their vigorous attack on physical determinism. Castaing had insisted that brain capacity, skin colour, and faces with protruding jaws (prognathic features) were not signs of arrested development. A well-directed education in a suitable milieu, he thought, could lead "the most decrepit race to an equal, even superior state" to that in which are found "many members belonging nominally to the most civilized nations." Whites in the tropics were not well off, while black artists in Paris could flourish.[56] In each

53 Manouvrier, "L'Ethnologie et l'Ethnographie dans l'Anthropologie," 163, 166–71; on this periodical of the scientific materialist faction, see Richard, "La Revue *L'Homme* de Gabriel Mortillet," 221–56; for Manouvrier in 1894, see "L'Anthropologie et le droit," 351, 365.

54 Manouvrier, "Rapports de la sociologie et de l'anthropologie," 877–8; for the intellectual and moral side to anthropology,.see "Le Classement universitaire de l'anthropologie," 75–109.

55 Hamy, preface, *Revue d'ethnographie* 1 (1882): ii; "De l'Origine et des progrès récents de l'Ethnologie," 37–47; for Hamy's career, see A.N. AJ 15 556.

56 Castaing, "L'Unité de l'Espèce humaine," 394, 397, 429 (citation); Edison, "The Society of Ethnography," 2; Castaing, "L'Ethnographie et la structure de la peau humaine," 133–4; Review of Eymar, "Les Peaux-Rouges," 89–90 on refusal to believe in essential

milieu, he claimed there could be "improvements" to the human forms that had adjusted to it.[57] At the 1878 Congress, he insisted, "civilization influences the spirit of primitive temperament." Ethnographers would find the causes of the prosperity or the degeneration of various peoples, and help those in need. Anthropology, in its circumscribed physical view, believed ethnography was only concerned with customs, but ethnography was quite capable of showing the modifications of types and languages through the general progress of humanity. History, which ethnography respected, would show the mixture of races, legends, and traditions, while anthropology disdained history and considered man nothing but an animal.[58] Castaing in these passages seems to come down solidly on the side of a cultural, environmental interpretation of the most important human attributes.

Like Castaing, Labarthe argued that all races were virtually equal, to be approached in a spirit of exchange and solidarity, law and fraternity.[59] Rosny at the outset had conceded that there might be separate human species, but this would make no difference since all would have "the Divine element itself, immaterial, immutable, eternal, and infinite."[60] He consistently doubted the importance of "osteological differences" compared to "emanations of intelligence."[61] He also supported the "virtual equality between races" since "inequality of races is morally revolting."[62]

The paradox is that the ethnographers' well-known disdain for craniology masked some of their own tendencies to continue to study physical aspects of human nature. This broad conception of ethnography, however repudiated by the founder Rosny, helps explain why anthropologists resented their rival society and could not unite with them as did colleagues in Britain or Germany.[63] In 1877 the artist Victor Dumas noted that natural man is well within the purview of ethnography since

inferiority and on Anglo-Saxons and French, 97–9; Castaing, "Sur le prognathisme," *Actes* 4 (1862–64): 201; at the 1878 Congress he castigated use of the "cephalic index" of long versus short-headed races, *Congrès international*, 119.

57 Castaing, "L'Ethnographie et la structure de la peau," 137.

58 Castaing, *Congrès international de l'ethnographie…1878*, 97–8, 299–301; see also environmental modification in *Ethnographie de la France*, 76.

59 Labarthe, "Esquisse d'un tableau préparatoire génésiaque," 146–7; Discussion sur le droit dans l'ethnographie, *Actes* 6 (1868–70): 115–16, 123.

60 Rosny, "L'Amérique," 108.

61 Rosny, "Rapport annuel. . .1863," *RO* 9 (1863–4): 366; also cited in Chailleu, "Léon de Rosny," 94.

62 Rosny, "De la Méthode ethnographique," 51–2, 54; "Eléments d'une ethnographie descriptive," 212, 215–16.

63 Stocking, "Qu'est-ce qui est en jeu dans un nom?".

ethnography is a philosophical, historical, and natural science.[64] Castaing, like so many psychologists and sociologists, stressed that the animal origins of human beings had to be studied as an introduction to ethnography. With one flourish he rejected the localization of brain faculties by the phrenologists Franz Joseph Gall and Johann Caspar Spurzheim. But he agreed with Broca that language faculties were related to a location in the brain and he used the divisions popularized by the very same Spurzheim to classify all faculties of the mind as affective, perceptive, and reflective.[65] In 1862, he reproved the critiques of phrenology by Adolphe Garnier and J.-P.-T. Flourens and hoped that his colleagues would "recognize in Spurzheim the true founder of the philosophy of the future."[66]

In 1888 Castaing even praised Broca's efforts at brain localization as a salutary kind of phrenology, a label Broca would have disdained for its unwelcome associations with the discredited Gall. Moreover, Castaing's estimate of women's intelligence, while less hereditarian than the extremist anthropologist Gustave Le Bon, accepted Le Bon's conclusion that the education of white European men differentiated them more from European women than were African men from African women.[67]

In a work of 1862 on Isidore Geoffroy Saint-Hilaire's classification of a separate human kingdom, Castaing insisted on the importance of both animal organs and an immaterial soul in understanding the human condition.[68] Unlike Quatrefages, he would not make religion and morality the criteria, since the Bushmen allegedly had neither. Animals also had reason and feeling. The key difference was that a human being manifested "a principle made in the image of God by organs appropriate to this destination," that is, by perception and intellect.[69]

While the ethnographers thus hesitated to repudiate the innate physical foundations of culture, they also cast doubt on their proclaimed egalitarianism by periodically reinforcing the racial or ethnic stereotypes of their era. They never acknowledged this contradiction and would have denied such prejudices if asked. Near the end of his life Castaing resubmitted his original program for ethnography dating back to 1860. The first section recognized the inseparability of "physiology and psychology"

64 Dumas cited in Zerilli, *Il Lato oscuro*, 12–13.

65 Castaing, *Classification de l'homme dans la nature*, (extrait de la *Revue orientale*), 45, 48, 54.

66 Castaing, *Classification de l'homme dans la nature*, 27.

67 Castaing, "Les Fonctions intellectuelles," 11–12, 22.

68 Castaing, *Classification*, 15–19.

69 Ibid., 41–3.

as an introduction to intellectual functions.[70] "Natural ethnography," a form of anthropology, the "science of type and species" was thus a necessary prelude to other forms of ethnography. As Paul Edison has remarked, in a reflection of tensions about immigration and national identity, Castaing also made a sensational comment about the need for an "ethnographic sense ... interested in the question of races so neglected in France, and for your misfortune, it's from that insane spirit of cosmopolitanism that has opened us up to anyone who wishes to come in, without even leaving us the option of destroying our enemies." This statement could be a distant forerunner of Herman Lebovics's notion of "True France" or of anti-immigrant concerns of the 1930s.[71]

Even the careful Rosny warned that the "black race," despite the higher civilization of the Senegalese Wolof, would relapse into barbarism without white guidance. He also accepted Isidore Geoffroy Saint-Hilaire's system of racial classification into somatic groups by skin colour, hair type, head shape, and the cephalic index of Anders Retzius (front-to-back head diameter divided by head width x 100).[72] Even in 1869, when he hoped ethnographers would study "nationalities," he agreed that there was a link between "physical characteristics and aptitudes." One had to take into account "some anthropological characteristics of races, inasmuch as these characteristics can influence the manner of understanding and realizing civilization."[73]

In 1888 Rosny worried enough about the physicalist trend to say that historical, economic, and psychological facts were primary to ethnography and that "societies, not races, undertake the work of progress and civilization."[74] And anthropologists would never accept his theory of the "méthode conscientielle" and "exactivism," a spontaneous judgment that is a pre-rational apperception and a vaunted sign of what Rosny called "spiritualist positivism."[75]

While Castaing and to an extent even Rosny had qualified the usual assumptions about the importance of nurture over nature, from the earliest years of the society other active members accepted the continued significance of membership in "racial" groups. Devout Christians were

70 Castaing, "Les Fonctions intellectuelles," *Bulletin de la Société d'ethnographie* 281–92; 7.

71 Castaing, *Congrès international. . .1878*, 460, repeated in *RO* 2nd series, 2 (1887): 286; Edison, "From Moral Brotherhood," 7–8; Lebovics, *True France*, 22; for debates on immigration in the late nineteenth century see Noiriel, *Immigration, antisémitisme, et racisme en France*, 136–205.

72 Rosny, "Eléments d'une ethnographie descriptive," 209–11; Isidore Geoffroy Saint-Hilaire, *Mémoires de la Société d'anthropologie de Paris* 1 (1860–63): 125–44.

73 Rosny, "Essai d'ethnographie théorique et appliquée," 7–8.

74 Rosny, "Conditions pour être ethnographe," *RO* 2nd series, 3 (1888): 207.

75 Rosny, *La Méthode conscientielle*, xii-xiii, 21, 29.

often partisans of human brotherhood and usually believed all men had souls and could not be equated with apes. However, the religious linguist Charles Schoebel found the "deformity of traits of peoples of Negro and Mongolic races was a transmission of the sign placed by God on the forehead of Cain to save him from the vengeance of his brothers." There was thus a "real separation between the white race and others." The Americanist Hyacinthe de Charencey questioned this characterization, but agreed that savages cannot achieve civilization on their own.[76]

In 1868 the instrument-maker of the Collège de France Joseph B. Silbermann heatedly contested the initial egalitarian position of the devoutly religious lawyer Eugène Minoret on the universality of legal principles. Minoret refused to acknowledge the existence of inferior races, since none failed to progress. Even aboriginal Australians had what they needed for their own civilization.[77] Silbermann, with the support of the anthropologist Royer, argued for "different laws for the different aptitudes of the races," while even Rosny agreed that the kingdom of Dahomey had no real laws. Silbermann insisted that Jews were never farmers or builders. Minoret, forced to concede the present inequality of some races, retorted, with support from Labarthe, that Europeans had no natural law permitting them to use cannons, opium, and brandy to exterminate others.[78] In the late 1870s Dr Antoine-Edouard Foley found the Polynesians of the Marquesas "quite indolent children."[79] The student of Icelandic literature Eugène de Beauvois also insisted that the "Eskimos" are an essentially stationary people fated to occupy lands where otherwise only polar bears can live.[80]

Contrary to scholars' usual claims, anatomical and physiological issues continued to intrigue the ethnographers. In 1864 Rosny himself called for an investigation of "principal anthropological characteristics," including crania and body parts.[81] The Chinese language professor at the Collège de France, Marie-Jean-Léon, marquis d'Hervey de Saint-Denys (1822–1899), president of the society in 1863–4, was responsible for the publication of an "anthropological photographic collection" from 1865 to 1867 with frontal, profile, and rear views of the bodies of

76 Charencey, "Le Déluge et les livres bibliques," 4; Schoebel, response to review of Charencey, *RO* 1 (1858–9): 404–5.

77 *Mémoires de la Société d'ethnographie*, 2nd series, 1 or 11 (1869): 264–5; Minoret, "Le Droit et l'ethnographie," *Actes* 6 (1867–70): 26.

78 Discussion, *Actes* 6 (1867–70): 113–28: see especially 113–14 for Silbermann and Minoret, 115 for Labarthe, 116, 122–3 for Royer, 117–18 for Silbermann.

79 Foley, *RO*, new series, 1 (1878): 309–10.

80 Beauvois, "Les Skraelings ancêtres des esquimaux," esp. 5–6.

81 Rosny, "Instructions ethnographiques," 28.

representatives of different peoples. The forty-two plates of nude figures included natives of Santo Domingo, south Africa, China, the Sudan, Moravia, a German Jewish girl, and a Samoyed. The military officer and explorer Louis-Emile Duhousset, a Société d'anthropologie correspondent, suggested a grid of measurement for photographs and drawings of individuals of different races – including the crania, lower jaw, and female pelvis so as to help arrive at average types of different races and perhaps cultural standards of beauty.[82]

In addition, there seemed to be special interest in the reproductive systems of both sexes even among the ethnographers. Labarthe became convinced that earlier onset of menstruation and nubility in Africans and Asians was primarily a racial trait, whatever the contribution of climatic and social influences.[83] One of the few questionnaires produced by the society was a detailed program in 1871 for investigation of the ethnic characteristics of reproductive systems of diverse races. A commission, including military officers Emile Calmette and Duhousset, as well as Labarthe, Rosny, and Clémence Royer, not only proposed to measure various body parts with the usual voyeuristic emphasis on women, but also to measure male genitals. They included questions on potency that had to assume an observer actually witnessing sexual acts. Hervey de Saint-Denys was so scandalized afterward that his name as president had been attached to the document that he formally disavowed it.[84]

The activities of the obstetrician Eugène Verrier (1824–1910), who was secretary-general of the society in 1885–88 and again in 1896–97, (he was a member of the Société d'anthropologie from 1883 to 1890 and secretary of the Société africaine de France from 1890 to 1895) demonstrated continued interest in physical anthropology, albeit with a bias toward the influence of the milieu akin to the evolutionary anthropologists.[85] In 1884, writing in Gabriel de Mortillet's journal of the scientific materialist anthropologists, *L'Homme*, he notified the anthropologists that pregnant women had easier labour "as one descends the ladder of civilized beings."[86] In 1887 he introduced to the Société d'ethnographie a memoir on Asian medical ethnography, largely concerned with physical characteristics.[87] However, in the final analysis, Verrier's interest in the physical did not automatically negate his belief

82 *Actes* 7 (1867–70): 169, 179–86.

83 Labarthe, "La Menstruation dans les races différentes," 70, 74.

84 Calmette et al., "Projet de questionnaire," 13–26, *Actes* 7 (1871–4): 66.

85 Sibeud, *Une Science impériale*, 316–18; for his Fourierism, which might have alienated the parti colonial, see Sibeud, "'Negrophilia,' 'Negrology' or 'Africanism'? 158.

86 Verrier, "De l'Obstétrique comparée dans les races humaines," 269, 272.

87 Verrier, "Ethnographie médicale," 39–51.

in the importance of milieu. In 1889 he focused on the physical milieu: how various soil compositions, climates, topography, air quality, and domesticated plants affected human survival. He also feared, like the degeneration theorists, that alcohol consumption could result in a "morbid heredity" in the urban population, so that rural migration to the cities might desirably be curtailed.[88] He thought Africans could benefit by further stimulation of their cerebral cells. A dabbler in the communal and feminist utopian socialism of Fourier, he supported the emancipation of women on the grounds that he had observed the equal capacities of women students. The structure of their brains, that is the amount of gray matter and number of circumvolutions, more than total weight and volume, were more important for their intellectual capacity.[89] For Verrier the analysis of women's brains did not inhibit support for entry of women into the public sphere.

The ethnographers from the 1870s onward would send very mixed messages on their attitudes to other races. The sensitivity to immigration beginning in the 1870s added social tensions to the discourse on hierarchy. The secretary-general of the society in 1873, Edouard Madier de Montjau (1816–1898), once a bank inspector in China and Japan, warned that Chinese immigrants, with their aptitude for commerce and good work habits, would be "materially redoubtable for Europeans" and a "cause of uneasiness for friends of peace, order, and ethics as we understand it in Europe." At the 1878 Congress Royer also feared the "overflow" of Chinese population in Europe.[90] While anti-Semitism, anti-immigrant sentiment, and fears of a "yellow peril" were conventional prejudices of the age, they do not fit well with proclamations of universal fraternity. Thus, cultural prejudices toward other "races," as well as anthropological allegations of natural inferiority, continued to affect those most inclined to boast of the influence of nurture over nature. As with the anthropologists, the view of gender roles was more flexible partly on the basis of alleged brain development, although the voyeurism of the questionnaire on reproductive systems objectified both intended female and male subjects.

A SCIENCE OF NATIONALITIES?

This apparent ethnocentrism certainly took place against a background of concern for the boundaries and justification of national identities.

88 *Congrès international des sciences ethnographiques ...* 1889, 13; Verrier, *De l'Influence des milieux dans la lutte pour l'existence*, 63–5.

89 Verrier, "Les Aptitudes de la femme et sa craniologie," 242–9.

90 Edouard Madier de Montjau, *De l'Emigration des Chinois,* 7, 9, 12–15 (citation on 15); Royer, *Congrès international ... 1878*, 163.

The French would continually question what nationality meant. Certainly there were large elements of agreement among both ethnographers and anthropologists that the peoples constituting nations were not equivalent to races and could be composed of several races.[91]

In 1864, as part of his huge program for ethnography, Labarthe urged the study of "ethnographic sociology," which would examine the economic and religious elements of a nation, the state, and the need of a "real nationality," not an "artificial" one, for independence and justice.[92] In 1864 Rosny had even tried to define ethnography as a "science of nationalities" and a nation as a "normal assemblage of individuals united by common sentiments, organization, traditions, history, languages, religious and civil customs, legislation, and art."[93] The next year Labarthe spoke of nations as subjective, and hoped for discussion of whether they were useful or harmful.[94] Castaing responded by questioning the focus on the nation, since it was an artificial ensemble, taking no account of natural divisions of races, or the historical growth of languages, or ethnic aptitudes.[95] Rosny however spoke of "normal nations" having a durable right to territory, while by implication "artificial" ones might not.[96] The common sense of culture was highly significant for this group, but they did not completely discard the sense of a pre-existing ethnic entity.

In 1869 Rosny reduced the significance of language groupings by noting that Semitic language speakers such as Jews and Arabs are more culturally distant than Jews and "neo-Latins." Moreover, overseas empires had no right to seize territory if they only needed trading outposts and fortresses.[97] In his 1897 presidential speech Rosny hoped that nationality would eventually disappear before international and unitary institutions. But, as Williams, Lebovics, and Edison have pointed out, he also affirmed in 1900 that "first nationality has to exist," before a later era of "fraternal association with fusion of national consciousness" and "no inferior peoples." Rosny added that "each people has a distinct and special

91 See Minoret in *Revue orientale* 6 (1868): 27 on multi-racial nations; For questions of nationhood relative to immigrants see Brubaker, *Citizenship and Nationhood*; and Noiriel, *The French Melting Pot*; *Etat, nation et immigration*; for the conservative implications of the nationalist position from the late nineteenth century to the Vichy regime, see Lebovics, *True France*.

92 Labarthe, "Esquisse d'un tableau préparatoire génésiaque," 154; Castaing, discussion in *Actes* 5 (1864–6): 162.

93 Rosny, *Rapport annuel . . . 1864*, 114.

94 Labarthe, "Esquisse d'un tableau préparatoire," 155–7.

95 Ibid., Castaing discussion, 162.

96 Rosny, "Essai d'ethnographie théorique et appliquée," 14, 16.

97 Ibid., 6, 14, 16.

place in the grand banquet of civilization," though this seemed to beg the question of virtual equality.[98]

Certainly at the time of foundation of the society, the nationality of the peoples of Romania was an important political issue. After 1871 the national identity of Alsace would be a recurrent topic for French scholars, particularly since Germans could argue that language was a key ingredient of nationality. Quatrefages refused to base national identity on ethnology. In an 1873 brochure Abel Hovelacque stressed that national identity was a "social rationality" resting above all on common interests. This era also saw the inculcation of the German language in ethnic Polish areas of Prussia. For Hovelacque, nations were composed of heterogeneous races and languages, but the "solidarity of united elements is justified by the will of the multitude."[99] He argued that basing nationality on language would somewhat increase French territory, but would lose the Basque and Breton areas, while Italy would find its area reduced.

Hovelacque certainly subscribed to racial hierarchy, but in Europe itself he believed in a multiplicity of races in each nation with no racial purity and no Aryan type. He could therefore claim that race difference was no basis for conflict, and that at least in Europe anthropology was conducive to lessening, not promoting, national rivalries. However, among the "least elevated layers of humanity" races, languages, and nationalities might coincide. Hovelacque believed colonialism was justified but probably not feasible because he doubted the "perfectibility" of inferior races.[100] In 1882 Ernest Renan delivered his classic lecture "What is a nation?" that declined to define a nation in terms of race, linguistic grouping, religion, or strategic interests. Nationhood to Renan was a "soul, a spiritual principle ... a daily plebiscite" that freed it from all ethnographic considerations.[101] Reviewed by Cyprien Issaurat for the anthropologists, Renan's essay appeared in the context of a spirited refusal to acknowledge the legitimacy of pan-Germanism and pan-Slavism. To Issaurat, "anthropology condemns pan-Germanism as much as pan-Slavism."[102]

98 Rosny, *Traité d'ethnographie théorique et descriptive,* 1: 32–4, 53–4; on nationality see Williams, "Anthropological Institutions," 339, cited also in Herman Lebovics, *True France,* 22.

99 Hovelacque, *Langues, races, nationalités,* (1873), 37; see also 9–20.

100 Hovelacque, *Langues, races, nationalités,* 2nd ed. 1875, 12–14, 33; Thulié, *1876–1906 L'Ecole d'anthropologie de Paris,* 23; Paligot, *La République raciale,* 51, 63; see chapter 3 on colonialism.

101 Renan, *Qu'est-ce qu'une nation?,* 26–7.

102 Issaurat, Review of Renan, 355–6; for the argument that even the controversy over Quatrefages's condemnation of the "Prussian" race as more barbaric ("Finno-Slavonic") than Germanic races was conducted in the context of separating nationality from race and therefore from inevitable racial conflict, see Chris Manias, "The *Race prussienne* Contro-versy."

After the turn of the century, anthropologist Georges Hervé, otherwise a hereditarian, stressed that "democracies are voluntary associations" that cannot immobilize people in anthropological frameworks of race or even of language. Germans themselves were not a single race, and there was a Celtic mixture in Alsace, rather than pure Germanic stock. The psychic temperament of Alsatians testified that psychic heredity developed by social milieu was independent of the ethnic heredity determined by characters of race (a line of reasoning Hervé would be unwilling to apply to Métis of "distant" races).[103]

With this general agreement on the significance of the nation, one would think that civic rather than ethnic nationalism would be prevalent, especially among the ethnographers. However Castaing betrayed the new ethnic variety of nationalism in 1878 when he argued that "sons of foreigners in France are not part of the nation," while the French abroad still are. He also offered the stereotypical opinion that Jews, though persecuted from all sides, preserved the "physical, moral, and intellectual characteristics of the race."[104] In 1889, twenty years after he had placed Jews as culturally closer to Europeans than Arabs, Rosny would separate the identity of language and people by saying, that, though Jews know the Hebrew language only in their prayers, they are still Semites.[105] In 1893 Rosny justified Japanese fear of foreigners in their home islands by saying, "all nations are threatened by the invasion of foreigners, the Western world by Jews, and America by the Chinese." Such casual remarks are perhaps as revealing as the programmatic statements of fraternity. The notorious anti-Semitic author Edouard Drumont, with his multifarious interests in racial theory, joined the society in 1892.[106] While neither the ethnographers nor the anthropologists were imbued with anti-Semitism, the ethnic components of nationalism were difficult to jettison.

Probably the summit of Rosny's contradiction of his usual egalitarian message was his concern in 1889 that lack of exemptions of the "enlightened" from military service was an unacceptable "blood tax." At that point he argued that "equality often produces evil," and that even anti-slavery agitation had been "more sentimental than rational." Moreover, in the world at large, "there are civilized and uncivilized nations" and "no fraction of humanity can hermetically close its borders to

103 Hervé, "L'Alsace," 288–9, 298–9; and 14 (1904): 296–7; Blanckaert, "La Crise de l'anthropométrie," 147.

104 *Congrès international … 1878*, 290–291.

105 *Congrès international … 1889*, 48.

106 Rosny, "L'Evolution politique et sociale du Japon contemporain," *Bulletin de la Société ethnographique* 2nd series, 7 (1893–4): 208, 216; membership list for 1892.

the currents of progress and civilization." Edison noticed Rosny's flirtation with eugenics, though Rosny drew back from the thought that the state could impose a decision on individuals not to marry or have children. If not for universal suffrage, Rosny even speculated one could tax the inactive and lazy. He hoped for voluntary restraint by the poor to limit family size so that "every being can serve his country and humanity in the place reserved by nature."[107]

Still society membership belies a simplistic explanation of xenophobia as the reason for this elitism. Even at the 1889 Congress, Haitian and Chinese military officers presented papers.[108] Aside from numerous Jewish scholars who were members, Rabbi Lazare Isidor, chief rabbi of the Central Consistory of France, became president of the Society in 1875.[109] Even more sensationally, in 1894 the affiliated African and Oriental Committee elected an all-Haitian slate of officers, including ex-probationary naval officer and law student Benito Sylvain as president (later Sylvain was a pan-African nationalist). Rosny's introduction of Sylvain once again endorsed the "virtual and natural equality of races" and castigated anthropologists for using "tools of the shoemaker" to assess "the relative value of different groups of humanity," thus providing an "excuse to crush entire populations and reduce others to the worst order of servitude."[110]

Sylvain himself consistently defended the perfectibility of the black race, contended that black Egyptians were the teachers of Europe, and urged a more humane French colonization, with progressive self-government, before switching to a mode of anti-colonialism.[111] Some scholars have suggested that this episode of electing Haitian committee officers could have alienated enough members to account for the reduction in adherents to the society.[112]

Rosny's last comprehensive treatise on theoretical and descriptive ethnography (1900–02) still maintained the superiority of "ambiantisme" (perhaps derived from the usage of "ambient" by Geoffroy Saint-Hilaire) and defended "equal rights for all peoples called to civilizations." Rosny, however, now seemed content to consign the savage to the natural

107 Rosny, "L'Idée de l'égalité," 17–18, 19–20, 24, 27 (citation); see Edison, "The Society of Ethnography," 1, on his brief flirtation with eugenics.

108 See Tcheng Ki-Toung, "La Nationalité chinoise, apercu ethnographique," 7–13.

109 *Actes* 8 (1874–77): 95.

110 *Bulletin de la Société d'ethnographie* 2nd series, 6 (1892–4): 243–4.

111 Ibid., 2nd series, 7 (1895–6): 20–23; Sylvain, *Du Sort des indigènes*, 236–43; Sibeud, *Une Science impériale*, 76.

112 Sibeud, "La Construction des savoirs africanistes en France," 225–7; on Sylvain see also Sibeud, "Negrophilia," 163.

historians of anthropology, while promoting for ethnography the study of the path to civilization. With lessening confidence in craniometry, Rosny thought ethnography needed autonomy from anthropology. Doubtless, his "pre-conscious reaction" of intuitive analysis would have separated him from orthodox proponents of scientific method.[113] But somewhat like Broca and the Muséum librarian Joseph Deniker, his final pronouncement would echo the sentiments of many philosophers and even Durkheimian sociologists: "race has meaning for the savages and barbarous, but little meaning any more for the civilized."[114] Hence, some cranial measurements would be useful for anthropologists, but "customs, institutions, and the influence of milieux are in a word sometimes as all-powerful on the physical modification of races as on their intellectual and moral transformations."[115]

ETHNOGRAPHY AND EMPIRE

One of the most striking similarities between ethnographers and anthropologists was their participation in a renewed imperial discourse justifying fitness to rule. Even if other races were perfectible, their evolutionary retardation meant that Europeans were needed to nudge them into progress, and European tutelage would be required for a long time before emancipation. The ethnographers voiced more spectacular anti-colonial pronouncements than the anthropologists ever did. But both groups adjusted their positions to a new associationist policy that would consider the particular traditions of indigenous peoples and never assume that emancipation was their birthright.

The ethnographers' *Revue orientale* began in 1859 with the proclamation of the "universalization of enlightenment," but assumed that Tunis "aspired to our civilization" and lauded the "rapid progress of French domination in Senegal." At a time when Madagascar was still contested among British and French missionary influences, the anonymous author commented "Madagascar has de facto never ceased being a French colony."[116] In 1862 the society proudly furnished the emperor with a geographical, linguistic, and historical *Tableau de la Cochinchine*, with several chapters by Rosny. Paul de Bourgoing, in praising the French influence

113 Rosny, *Traité*, 1: 25–26, 45–6, 50; 2:172; Zerilli, *Il Lato oscuro*, 77–8.

114 Rosny, *Traité*, 2: 194.

115 Ibid., 2: 196, 212.

116 Rosny, "Amérique et Orient" *Revue orientale et américaine* 1: 1859, 1–3, cited in Chailleu, "La Revue orientale," 95; 91 for prospectus; see Charencey's call for Christian influence in Tunis in Review of Dunant, "Notice sur la Régence de Tunis," *Revue orientale et américaine* 1 (1859): 291.

on an "admirable colony," was pleased that the French army would not encounter "fanatical enthusiasm here, since Buddhism is not hostile to us." The geographer Eugène Cortambert (1805–1881) contributed stereotypical images of the proud but filthy Indochinese and the fearful "big children" in Cambodia.[117]

The Christian component of this imperial discourse was also clear in Castaing's contention that French "moral force ... for the honor of the Christian name" was needed to reform Indochina and his praise of French colonists in Algeria, who brought "Christianity and civilization" and equality of rights while respecting local customs.[118] Labarthe similarly expected the colonization of Madagascar to spread the gospel and to regenerate the Orient as well as France itself.[119] Alexandre Bonneau (1820–?), a geographer, society councillor, and journalist for *L'Opinion nationale*, uttered the common refrain "we have the right to dominate by our civilization," since "not all races and climates can produce civilization." The only obligation would be to "elevate the conquered to a new intellectual and moral level." Europeans should also try to improve Haitian blacks to the "as yet unknown limits of their aptitude."[120] Verrier, speaking of Africans, also favoured a gentle "moral and peaceful conquest" so that the "disinherited races would be called to share the benefits of civilization and one day the direction of the world."[121]

However, the ethnographers displayed fits of humanitarian anti-imperial sentiment. In 1869 they discussed a series of resolutions concerning the justice of colonization and what they called "ethnographic law." The first concerned whether the appropriation of an entire continent by a colonizing nation after merely planting its flag there, often approved by custom, was legally justifiable. The resolution proposed that property seized be limited to the "territories necessary for the subsequent development of colonies," while other territories would be open to all enterprises. The target was the English monopoly on Australia. Some members repeated Rosny's claim that trading posts and fortresses would often suffice for European interests. Thirty-three members voted

117 Bourgoing, ed. *Tableau de la Cochinchine*, iv, viii, xi, 129; see also Société d'Ethnographie *Rapport annuel 1864*, 69; and Rosny's commitment to Annam in *Traité*, 2:157.

118 Castaing, "Sur la question marocaine," *Revue orientale et américaine* 3 (1859–60): 17; Castaing, Review of Eymar, "Les Peaux-rouges," 97–99; Castaing's reconciliation of the Bible with archaeology earned the ire of Gabriel de Mortillet in *L'Homme* 2 (1885): 153–6.

119 Labarthe, "De la Colonisation de Madagascar," 299–300.

120 Bonneau, "Le Droit de conquête," 372, 374–5; *Haiti, ses progrès, son avenir*, 97–8.

121 Verrier, "Les Races noires de l'Afrique," 121–2, 127–8.

in favor, including Claude Bernard, Rosny, Minoret, Castaing, and Labarthe, while only four voted against, and four abstained.

The second resolution affirmed that law prohibits all "violent dispossession of indigenous sedentary or nomad populations, even if the invaders could better exploit the land worked by the labour and sweat of the first proprietors." Twenty-seven, including Silbermann and Verrier, as well as Minoret, voted in favour, with only two opposed (including the rather ferocious Clémence Royer), but eight abstained (including Rosny, Castaing, and Bernard). Hence Rosny himself had doubts about prohibiting Europeans from developing otherwise unexploited resources.

The third resolution provided that "both law and morality repulse those pretended contracts, by which civilized Christian nations usurp from credulous peoples vast rich countries for an imaginary price." Twenty-seven, including Rosny, voted for this resolution, four voted against, including Royer and Schoebel, and six abstained. The fourth resolution adopted unanimously approved the sentiment that "justice no less reproves the pact by which civilized nations exchange among themselves immense territories and trade the populations living there." Such an opinion would have condemned exchanges and agreements as took place later at the Congress of Berlin in 1878, although there is no record of the reaction of the society at that time. The fifth provided that "cessions, exchanges, rectifications of borders can only be legitimized by the consent of the populations." Thirty voted in favour of this, one abstained, and three opposed (including Rosny and Verrier). Rosny and Verrier seemed thus to doubt whether indigenous populations had the capacity to determine their own destiny. The sixth resolution stated that "indemnities and reparations demanded of peoples not be degrading or ruinous." Thirty-eight supported this, while Rosny was one of two who abstained. Finally, as if retrospectively condemning the French pretext for the Algiers expedition of 1830 or the more recent Mexican expedition, a resolution stated that the "pretext of avenging an offense, and of safeguarding interests, never authorizes an attack on the independence of a foreign people, or modifying by force its constitution and laws." Twenty-seven voted in favour, while Rosny and two others abstained. The architect and archeologist Charles Geslin claimed that the votes against or abstentions were only protests against the drafts deemed to be incomplete, obscure, or insufficient, rather than against the principles endorsed by the majority. However, Rosny abstained on an approved resolution not to attack the independence of any foreign people.[122]

122 Discussion in *Actes* 6 (1867–70): 191–7.

The turn toward associationist discourse was apparent at the International Congress of Ethnography in 1889[123] There the Appeals Court lawyer Célestin Pret helped obtain endorsement for a resolution commenting on British rule in India. The initiative came from the Belgian linguist Eugène comte Goblet d'Alviella and the physician and Islamic scholar based in England, Gottlieb Wilhelm Leitner. The resolution stated that Europeans should "attempt to develop the culture of [Asian] populations, in conformity with their own genius, their ethnic traditions, and particular languages." [124] Around the same time, Verrier had suggested that, while France in Africa needed to exterminate slavery, cannibalism, human sacrifice, and mistreatment of women, the ideal vehicle would be the protectorate, as in Tunisia, Madagascar, and the Congo. He praised the Congo administration of the explorer Pierre-Paul-François-Camille Savorgnan de Brazza, who was known for his concern for native welfare, and suggested that European trade would benefit subject populations.[125]

CONCLUSION

The ethnographers in principle always affirmed the superiority of nurture over nature. But they illustrate the difficulty of complete gravitation toward one pole of the debate. For a group of students of culture, they refused to surrender their interest in physical investigations or their belief that race matters. They praised some forms of phrenology, suggested that Africans needed guidance, and in Rosny's case even intimated that egalitarianism was misconceived. And while they apparently risked the very existence of their society on the integration of Haitians, their casual comments on Asians and Jews as well as their flirtations with social elitism and ethnic nationalism seemed to mock the message of universal brotherhood and virtual racial equality. They subtly approved racial hierarchy while nominally condemning it.

Meanwhile, the anthropologists tended either to limit the definition of ethnography to a descriptive study or to see it as an auxiliary branch of the superior study of anthropology. They never gave up the terrain of the cultural, while many of them insisted it was secondary. Quatrefages and Manouvrier, however, were among the most notable creators of a

123 On associationism see Betts, *Assimilation and Association*; Conklin, *A Mission to Civilize* sees the turn to associationism happening after World War I.

124 *Congrès international des sciences ethnographiques ... 1889*, 55.

125 Verrier as President of the Société africaine, *Bulletin de la Société d'ethnographie* 2nd, 2 (1888): 214–18.

countercurrent that genuinely wished to introduce a sociological dimension to anthropology to complement head measurement.

Even those who highlighted milieu over heredity did not generally lose their enthusiasm for empire, albeit one humanely administered. Despite their apparently anti-imperialist resolutions, the ethnographers had still never abandoned the idea that certain peoples are retarded, and therefore in need of guidance. The anthropologists would also embrace this viewpoint. But there would be considerable variation on the subject of empire, partly because some anthropologists would consider it not feasible or not worthwhile because of the very hostility of the milieu and the alleged retardation of subject peoples.

The Société d'ethnographie would lapse into inactivity after the turn of the century, until the ventures of folklorist Arnold van Gennep and interest from the Colonial Office would revive ethnographic institutions in the period from 1910 to 1913. Their fitful initiatives form a fascinating contrast to the more renowned and institutionally successful saga of the anthropologists.

3

The Anthropologists

THE LITERATURE ON the Société d'anthropologie has been far more abundant than on the Société d'ethnographie. Previous analysis has discussed the positivist and materialist factions among the anthropologists, the emerging influence of Darwin and transformism, aspects of racial theory, and the anthropologists' institutional success.[1] These issues need no detailed reconsideration, but fruitful exploration of the nature-nurture debate in the periodical publications of the society and independent anthropology journals is still possible.

The anthropologists' discriminatory reading of the influence of the milieu and its complicated relationship to heredity needs further attention. We may also separate mainstream anthropologists who admitted that some progress through the social milieu was possible even for non-Europeans from the most extreme hereditarians who practically nullified the influence of milieu. To magnify the influence of heredity alone meant showing no qualms about unbridled domination of allegedly inferior races who could never improve. In addition, the anthropological countercurrent that reintroduced sociological considerations needs re-evaluation for its strength in criticizing head measurements and its limitations in failing to surrender essentialism. Many subtleties in the anthropologists' positions on gender issues made them partisans of a feminine "nature," even when some sympathized with women's emancipation and deplored mistreatment of women outside of Europe. In the last analysis, with a few exceptions, they considered the social roles of women so important that even the educability or increased brain size of European women would not automatically assure

1 References previously noted to Williams, "The Science of Man"; Harvey "Races Specified, Evolution Transformed"; Blanckaert, ed. *Les Politiques de l'anthropologie*; Hecht, *The End of the Soul;* Wartelle, "La Société d'Anthropologie de Paris"; Paligot, *La République raciale.*

an endorsement for equality in the family and society because of the desire for social stability.

Finally, the debate on imperialism suggests that for a time those believing in the unity of the species could be enthusiasts for empire, while those who were most prejudiced against other races had reasons to doubt the feasibility of European domination in tropical climates or to worry about racial mixing. There was an anti-imperialist and humanitarian minority voice among the anthropologists as among the ethnographers. Yet the majority ultimately desired a "scientific" colonialism that would apply anthropological and ethnographic knowledge about racial crossing and assure French domination while adjusting to the customs of each people.[2]

The Bulletins de la Société d'anthropologie de Paris (beginning in 1859) formed the basic record of the society, while more extensive articles appeared in Paul Broca's *Revue d'anthropologie* (1872–89). This journal, edited after Broca's death by his disciple Paul Topinard, merged with the ethnographer Ernest-Théodore Hamy's struggling *Revue d'ethnographie* (1882–89) and an archaeological journal edited by Emile Cartailhac to form the independent *L'Anthropologie* (1890). A brief flurry of activity by the scientific materialist faction produced *L'Homme* (1884–87), while the professors at the Ecole d'anthropologie felt the need in 1891 to establish their own reflection of the curriculum in the *Revue mensuelle de l'école d'anthropologie.*

LAMARCKIANISM AND THE INTERTWINING OF HEREDITY AND MILIEU

Previous literature has insufficiently emphasized the discriminatory application of Lamarckianism in the society. The majority of members in the 1880s balanced the idea of a fixed human nature with adaptability to the milieu and transmissible acquired characteristics.[3] The ultimate derivation from the naturalist Jean-Baptiste Lamarck stemmed from his argument that "new circumstances, having become permanent for a race of animals, give new habits to them ... one part will be used preferentially to another, and in certain cases, there will be the total lack of the part

2 Blanckaert, *Politiques de l'anthropologie*, 98, 113–24, on acclimation of Europeans being less of a deterrent in debates in the 1870s and 80s than in the early years; Paligot, *La République raciale*, 51–77.

3 Broca, "Anthropologie," in *Mémoires d'anthropologie*, 39–40; Blanckaert, "L'Anthropologie lamarckienne à la fin du XIXe siècle," 611–29.

that became useless."[4] For Lamarck, organisms in constant transformation modified structures by active responses to a new environment, and eventually accumulated adaptive improvements.[5]

Anthropological Lamarckianism was double-edged. Heredity and milieu were so intertwined that it became almost impossible to oppose them. Indeed as the physician and physiologist Charles Richet reported to the Société d'anthropologie on "psychological heredity" in 1884, "heredity and milieu are not antagonistic, but synergistic."[6] But the argument then as now was about their respective contributions. If heredity were all-important, or operated more powerfully now than in previous eras (as was argued by Gustave Le Bon for human races), the racial divisions would be more rigid and possibly even insurmountable to the point of questioning the unity of the species. If milieu were all-powerful, an optimistic view of human progress was built into natural processes, so that nature transmitted what nurture accomplished. Or alternatively the milieu might operate differentially, depending on the inherent nature of races. This latter view was the discriminatory paradigm. Lamarckian anthropologists, whether socialists or conservatives, generally believed that civilized people were adaptable, while culturally backward "savages" or those from "inferior races" were not or could only gradually change.[7] The variations of this paradigm could determine whether and how quickly other peoples could absorb European civilization. The consequences could affect the manner of colonization or domination.

THE HEREDITARIANS: BROCA, TOPINARD, AND THE DOMINANCE OF NATURE

The founder and first secretary-general of the society, the neurologist Paul Broca, illustrated the hereditarian inclination so prevalent among the anthropologists, though Broca himself was only a reluctant supporter of transformism. His successors would more fully adopt a differential Lamarckianism that favoured Europeans and disadvantaged all others.[8]

4 Lamarck, *Philosophie zoologique*, 221–2 and generally 218–51; on Lamarck see Burkhardt, *The Spirit of System*; Barthélemy-Madaule, *Lamarck, the mythical precursor*; Laurent, ed. *Jean-Baptiste Lamarck*.

5 Wolfgang Lefèvre, "Inheritance of Acquired Characters," 100–1.

6 Archives de la Société d'anthropologie, 1K3, 6 mars 1884; Dias, *La Mesure des sens*, 232.

7 For the "racial evolutionist paradigm," see Mucchielli, "Sociologie versus anthropologie raciale," 78–81.

8 On the early belief in fixed species and the limits of Broca's belief in transformation even in the Lamarckian sense, see Blanckaert, "La 'Nuit des origines'," Broca juge de Darwin," *De la Race à l'évolution*, 319–57.

Broca also manifested the compatibility of agnosticism and republicanism with full-fledged belief in racial hierarchy. While he avoided politics in the tense early years of the society, he ended his life as a committed republican senator (in 1880) through the patronage of the Opportunist Republican faction leader Léon Gambetta. Broca resolutely supported secular schooling and was modestly progressive on gender issues with support for secondary education for girls (albeit oriented to traditional family roles) and the award of a medical degree to a woman.[9]

Progressive politics of the republican French variety however could mesh with a commitment to differentiating human beings by physical features that allegedly correlated with intellect and character. Broca supported the "high importance of craniology ... because one hopes to find there some information relative to the diverse intellectual value of human races."[10] He argued that "innate and very different aptitudes" of Australians, as contrasted, for example, with Hawaiians, would "resist all efforts at civilization, even those pursued with the most intelligence and patience."[11]

Broca inclined toward polygenism, that is, the separate creation and development of each "race" (implying that each race was a separate species without settling the issue). In the early years of his interest in anthropology (1858–65) he was even a believer in fixed races.[12] In his memoir on hybridity, Broca claimed that fertility and general health among humans would vary with the degree of proximity of races. "Eugenesic" races would become stronger and remain fertile by mixing, while "paragenesic" (partially fertile) or "dysgenic" (infertile) crossing could not produce viable offspring.[13] In the ethnology of France, Broca showed that there was a mixture of long-headed and wide-headed types, tall blondes as well as short, brown-haired stock.[14] Contrary to the doctrines of racial theorist Joseph-Arthur, comte de Gobineau, who postulated decline (including the Latin races) with any loss of racial purity, this racial mixing could be beneficial and strengthen advanced civilized peoples. Somatic types

9 Wartelle, "La Société d'anthropologie de Paris," 134–5.

10 Broca, "Sur le Volume et la forme du cerveau," 139, 198–9; Schiller, *Paul Broca*; Blanckaert, "Préface," *Mémoires d'anthropologie*, i-xliii; Gould, *The Mismeasure of Man*, 73–113.

11 Broca, "Sur le Volume," 198–9; for discussion of the controversy between Broca, who advocated the importance of brain volume and weight, and Pierre Gratiolet, who privileged vital forces and instincts to arrive at his own racial hierarchy, see Blanckaert,"La Mesure de l'intelligence," *De la Race à l'évolution*, 143–74.

12 Blanckaert, "La 'Nuit des origines'," *De la Race à l'évolution*, 319–57, esp. 322–4.

13 Broca, *On the Phenomena of Hybridity*, 21–4, 33–60; Blanckaert, "Of Monstrous Métis," 42–72, and *De la Race à l'évolution*, 89–90.

14 Broca, "Recherches sur l'ethnologie de la France," 6–15.

and racial differentiation were also key issues, as Blanckaert has shown, for some political questions such as minimum height for military conscription and the ability of Europeans to acclimate in the colonies.[15]

As was fitting for a political progressive, Broca favoured belief in the perfectibility of Europeans. Contrary to the competitive applications of Darwinism advocated by Darwin's French translator Clémence Royer, he even advocated "social selection" by which "society owes protection to all its members [even the weak] … one has to accept this principle or return to the savage state." Civilization, while decreasing the average brain weight apparently due to less immediate challenges to most people, increased the number of exceptional individuals with very high brain weight and also increased the weight of the all-important anterior cranium that brought forth progress. Therefore, education and "equitable institutions" can "permit each individual to obtain a position proportional to his utility."[16] Broca favoured both open, unprivileged competition for employment and, with a hereditarian twist, voluntary planning of marriages of the best endowed. But he was skeptical that those outside of the hereditary influence of "enlightened families" could ever take advantage of what education can offer.[17]

Paul Topinard (1830–1911), the secretary-general of the society from 1880 to 1886, was the most enthusiastic mainstream heir of Broca's infatuation with craniology.[18] More politically moderate than the socialists among the society's materialist faction, he was a conservative republican with echoes of John Stuart Mill's elitism. Topinard's ideal society would

15 Blanckaert, "La Crise de l'Anthropométrie," 111–14; Blanckaert also distinguishes a faction in the society of extreme hereditarians influenced by or working in parallel with the reasoning of Gobineau on the preference for racial purity – including Eugène Dally, Clémence Royer, Jean-Christian-Marc-François-Joseph Boudin, Joanny-Napoléon Périer, and Anatole Roujou; see *De la Race à l'évolution*, 100–1.

16 Broca, "Les Sélections," from *Revue d'anthropologie* (*RA*) (1872) reprinted in Broca, *Mémoires d'anthropologie*, 244–5, 248–9 (vol. 3 in original); for perfectibility of various races and the non-cosmopolitanism of Georges Pouchet and Jean-Christian-Marc Boudin, see *Bulletins de la Société d'anthropologie de Paris* 1 (1859–60), 338, 347–50, 358–9, 375–9, 389; for Pruner-Bey, 493–4; see also Blanckaert, "La naturalisation de l'homme," 21; Williams, "The Science of Man," 92–5.

17 Blanckaert, *De la Race à l'évolution*, 356–7.

18 On Topinard's dispute with the scientific materialist faction, see the press documentation in Archives nationales F17 13558 Ecole d'Anthropologie; Muséum MS. 2256, Correspondance de Hamy, Topinard's letter of 4 november 1886 where he speaks of unmasking his opponents and still wants votes to counteract the "scientific materialist" group; Wartelle, "La Société d'anthropologie de Paris," 150; Harvey, "Races Specified," on the Topinard vs Mortillet group quarrel, 68–76; Williams, "The Science of Man," 193–7; see Topinard's self-defence in Topinard, *La Société, L'école et le laboratoire*; for Jules Fauvelle's indictment of his mismanagement of funds, his "deplorable misconduct and insatiable vanity" (a statement not to be read at a general meeting) see Archives de la Société d'anthropologie, Registre des procès-verbaux du comité central, 10 juillet 1890.

provide public education for all and protection for abandoned children, but a diluted democracy with indirect voting or alternatively plural votes for the wealthy.[19]

Despite later agreeing that racial divisions were artificial constructs, Topinard maintained the importance of classifying nineteen persistent basic human types, based on apparently fixed hair type, skin colour, eye colour, cranial form, and shape of face and nose. He thought four of these might be different species. Later he called races "constant varieties" of a single species without resolving the monogenist-polygenist debate.[20]

Brain anatomy was still critical for "consideration of moral and material production of the origins of peoples and their faculties." Since mixed peoples alone were observable, the critical task of anthropologists was to dissociate them into races. There was a need to "determine the elements and establish the hierarchy of them [the pure types]."[21] Despite this hereditarian bent, Topinard was not so extreme as to ignore that milieu, through conditions of life and healthfulness, would affect even inheritable traits such as height.[22] Unlike the nationalist conservatives, he also avoided racial anti-Semitic thought with the observation that Jews were too diverse to be a race.[23]

Topinard's beliefs exemplify that differential Lamarckianism in which Europeans, but not inferior races, could inherit moral habits. He felt free to speculate about how moral habits of solidarity, charity, and, altruism might be inherited and, after twenty generations, might become instinctive.[24] Yet in contrast even to prehistoric Europeans, "yellow races" have "little initiative and are subject to immobilization" while "all black races have an incapacity to elevate themselves on their own."[25] He believed that Africans could advance, but probably not without the incentive of the slave master.[26] Yet he refused to recognize this conclusion as a prejudice. Referring to American attitudes denigrating blacks, he dared claim that "the anthropological notion of race is absolutely foreign to these prejudices."[27]

19 Topinard, *Science et foi*, 424–7, 552–60.

20 Topinard, *L'Homme dans la nature*, 17–21.

21 Topinard, *Eléments d'anthropologie générale*, 166–7, 194.

22 Topinard, "Etude sur la taille," 83.

23 Topinard, review of Bernhard Blechmann [misprinted as Betchmann], "Ein Beitrag zur Anthropologie der Juden," *BSAP* 3rd series, 6 (1883): 84.

24 Topinard, "Conclusions et applications de l'anthropologie," 663, 690; for his critique of August Weismann, see Topinard, Review of Alfred Russel Wallace, "Le Darwinisme," 326–31, and Review of Weismann, "Essai sur l'hérédité et la selection naturelle," 331–6.

25 Topinard, *Science et foi*, 175, 227, 348, 365.

26 Topinard, "De la Notion de race en anthropologie," 643–4.

27 Topinard, *L'Anthropologie* 1 (1890): 382; for his elitist political views, see *Science et foi*, 392, 424–7, 432, 469, 489 on wealth tax, 499–500 on voting, and 545–6n. on Jews and

Topinard also refused to give up his Lamarckianism in the face of Darwin's natural selection and August Weismann's theory of continuity of germ plasm. He thought Darwinists vulnerable on explaining the causes of variation, while he recognized that Lamarckians had difficulty explaining why heredity produced constancy, and why the milieu did not act more decisively. He was obviously one of the more moderate Lamarckians in not being overly optimistic about the potentialities of human nature even among Europeans, but Lamarckianism did not inhibit his belief in racial hierarchy.[28]

The society however had a collection of hard-line hereditarians, such as the lawyer, linguist, materialist, and radical party politician, Abel Hovelacque, director of the École d'anthropologie from 1890 to 1896.[29] At the 1878 international congress of anthropology, Hovelacque asserted one could tell "superior and inferior races" by physical features, but also by "moral" (cultural) attributes such as "absence of good treatment of women, and fetishism."[30]

In a popularized survey of anthropology co-authored with the physician Georges Hervé, Hovelacque insisted "some races are closer to the apes," and that the now discredited facial angle of Peter Camper was still a good approximation for determining their place. They also believed that only migrations and interbreeding could serve as explanatory factors for the existence of human races, since Darwin had successfully demonstrated that the "distinct characteristics of human races cannot possibly be explained by milieux, climate, food, and lifestyle."[31] Like some early polygenists, the authors predicted the limits to acclimation of all races in unfamiliar climates – a view that paradoxically could affect the feasibility, though not the justification, of imperialism.[32]

Similarly Moindron-Sigismond Zaborowski (1860–1928) in his 1892 Broca lecture on the "Differences and Future of Human Races," spoke of the "mental disparities that cannot be eliminated" between the superior and inferior races.[33] Yet Zaborowski, like Broca, thought that education, diet, and social refinement could increase the resemblance of uncivilized

the Church; for Manouvrier's hostile comment on a "science extracted from the brain with teaspoons," see Manouvrier, Review of Topinard, "Science et foi," 210–12.

28 Topinard, Review of Weismann, 332–6.

29 Desmet, "Abel Hovelacque et l'école de linguistique naturaliste," 55–93.

30 Broca, "Congrès international des sciences anthopologiques," 752.

31 Hovelacque and Hervé, *Précis d'anthropologie*, viii, 70, 74, 193 (citation), 214, 257; on difficulty of racial classification, 222.

32 For the debate on "Hottentots" and "Bushmen" see Mucchielli, "Autour des 'Instructions aux voyageurs sur les Boschimans' d'Henri Thulié," 201–42.

33 Zaborowski, "Disparité et avenir des races humaines," 38–9.

to civilized facial structure.[34] He, too, would later express doubts about the practicability of imperialism but would also question the ideological reasons pretending to benevolence rather than simply admitting it was a strategic necessity.

Clémence Royer, a member of both the Société d'ethnographie and the Société d'anthropologie (in 1870 the first woman to be accepted by the anthropologists) expressed a sort of society vulgate opinion on phenomena of heredity, although her acid critique of Darwin's pangenesis theorem of particles collected from the entire organism in the germ cells was thoroughly personal. Her "dynamogenesis" theory allowed that the milieu through intrauterine accidents and external physico-chemical forces could affect heredity and produce atavisms. Heredity was a conflict of forces of preservation and of atavistic return to earlier forms variable in different environments. The advanced races would inherit accumulated intellectual skills, despite some regressions and the danger of freezing hereditary aptitudes. She remained, however, among the most skeptical about the capability of non-Europeans for improvement.[35]

HEREDITARIAN EXTREMISTS: LE BON AND VACHER DE LAPOUGE

The craniology of Broca, Topinard, and the hereditarians had the potential to turn into the most virulent exposition of racial hierarchy by anti-republican authors. Paligot has argued for the distinction of two streams in French anthropology, the nationalist and anti-Semitic conservatives (or maverick anti-Marxist socialists who became radicals of the right), and the progressive republicans, who did not endorse nationalism, racial purity among Europeans, or anti-Semitism.[36] The nationalist conservatives were certainly the most likely to stress indelible heredity with little prospect of overcoming the taint of inferior nature. Gustave Le Bon (1841–1931) and Georges Vacher de Lapouge (1854–1936) achieved considerable notoriety, though neither was in the anthropological mainstream. Le Bon, a political conservative, received the coveted Godard prize from the society in 1879 and published in their periodicals until 1884. He resigned with great irritation when Topinard refused to accept a memoir on geography in the *Revue d'anthropologie.* Afterward, Le Bon

34 Zaborowski, "Hérédité. II. Anthropologie," 1169–70.

35 Royer, "Deux hypothèses sur l'hérédité," 479, 673–4, 685; Cyprien Issaurat, Review of Clémence Royer, 330–41; on Royer see Blanckaert, "L'Anthropologie au féminin," 23–38; Blanckaert, "'Les Bas-fonds de la science française'",124; Harvey, "*Almost a Man of Genius,*" 95–6, 99, 108–9, 135–6.

36 Paligot, *La République raciale,* 89.

wrote, "Since that time I have renounced giving anything to the society."[37] His prizewinning study purportedly demonstrated that larger average brain capacities existed in white races, men rather than women, nobles rather than common people, and civilized rather than primitive groups. Even more important, the "advanced" races had more individuals with superior-sized crania, who could develop human progress.[38]

Le Bon later argued that "slow change produced by variability of milieux, selection, and accumulated by heredity" differentiated the races. The milieu then became too weak to modify the fixed racial characteristics. Considering Jews as a race, he observed that the weak milieu is why "sons of Israel in every latitude preserve their invariable type."[39] Still, modern races were matters of character and intelligence, not just physical endowment. But Le Bon then proceeded to nullify the importance of milieu in the present by arguing that "historical races" had a "mental constitution as fixed as an anatomical constitution," as if they were biological races. Most individuals in historical races would have common sentiments, needs, customs, and traditions reproduced as a "hereditary aggregate." Race determined the way peoples reacted to events and to changes in the milieu.[40] Le Bon's opening up of the definition of race to include cultural characteristics would later lead to an invidious labelling of people based on culture rather than physical attributes. The consequences would become visible in the period of the 1930s and during the Vichy regime.

Topinard allowed the other notorious extreme hereditarian, Vacher de Lapouge, a onetime socialist and then radical of the right, to use the *Revue d'anthropologie* as a platform for his contention that "in the next century peoples will slit each other's throats by the thousands for the sake of one or two degrees more or less of the cephalic index [ratio of anterior-posterior length of head to width of head]."[41] Topinard also published excerpts from Lapouge's courses at the Montpellier Faculty of Sciences in 1888. Inspired by the British founder of eugenics Francis Galton, Lapouge would recommend that prizes and bonuses should be given to the elite to hasten evolution. Ideally the inferior would be

37 Archives de la SAP inventory, 1L1 and 1K2 for Le Bon's submissions; correspondance, E, letter of July 26, 1888; Le Bon, "Notes sur l'étude du caractère," 497, 499; on Le Bon see Nye, *The origins of crowd psychology*.

38 Le Bon, "Recherches anatomiques et mathématiques," *RA* 8 (1879), 101–4.

39 Le Bon, "Applications de la psychologie à la classification des races," 594.

40 Le Bon, *Lois psychologiques de l'évolution des peuples*, 2–3; Paulhan, Review of Le Bon, "Les Lois psychologiques de l'évolution des peuples," 513–4, 516, 519.

41 Vacher de Lapouge, "L'Anthropologie et la science politique," 151; Hecht, "Vacher de Lapouge and the Rise of Nazi Science," 285–304, cited on 287; Quinlan, "The Racial Imagery of Degeneration," 393–413.

eliminated and superior individuals would be propagated for improvement of the species.[42]

Opposing "sentimental egalitarianism," Lapouge revelled in cranial measurements allegedly showing that the brave, enterprising dolichocephalic (long-headed) Aryans (such as the old French nobility and the tall and blond in every civilization), now thought to originate not in India but in northern Europe, would forever lead the plodding, peasant-type brachycephalics (wider heads).[43] Eugenic action was imperative to ensure the victory of the dolichocephalics, to counteract the regressive selection of wars where the bravest died, and to discourage the peasants. He believed that the fusion of the two types had led to reduced vitality and a lower birth rate.[44]

Non-European races were "incapable of being civilized," as revealed by Haiti, other islands in the West Indies, and Sierra Leone. Moreover, the truly primitive Fuegians, Australians, and Bushmen could not become cultivated or make discoveries.[45] As for Europeans, Lapouge was pessimistic about the projected triumph of the brachycephalics. Educating them without eugenic measures would develop and create nothing, and risk the breeding of degenerates.[46] Lapouge had limited success in France because, like Gobineau before him, he satisfied neither the more devoutly religious and peace-loving spiritualists nor the narrower French nationalists or Celtic supremacists uninterested in Aryan racial theory.[47] By 1906 a reviewer in *L'Anthropologie*, with exceptionally poor prophetic acumen, lacerated Lapouge: "scientifically, the Aryan is dead, Monsieur de Lapouge cannot revive it."[48] Regardless of events in Germany, the era of the 1930s would demonstrate the revival of Lapouge.

42 Vacher de Lapouge, "Sur l'Hérédité," 516–7; "De l'Inégalité parmi les hommes," 9–10.

43 On Montpellier crania, see Vacher de Lapouge, "Crânes préhistoriques de Larzac," 681–95, and "Crânes de gentilshommes et crânes de paysans," 317–22; on egalitarianism, see *Revue d'anthropologie* 3rd series, 2 (1887): 145; see also "Questions aryennes," 181–93. The German anthropologists and linguists Theodore Poesche and Karl Penka in 1878 and 1883 were the originators of the idea that Aryans came from the Nordic countries.

44 Vacher de Lapouge, "La Dépopulation de la France," 73–4.

45 Vacher de Lapouge, "De l'Inégalité parmi les hommes," 11–12.

46 Vacher de Lapouge, "L'Hérédité dans la science politique," 177–8, 187.

47 Massin, "L'anthropologie raciale comme fondement de la science politique," 269–334.

48 Delisle, Review of Emile Houzé, 698–700.

LETOURNEAU: RACE TRUMPS SOCIOLOGY

Topinard's materialist and Fourierist socialist colleague, the physician Charles Letourneau (1831–1902), replaced him as secretary-general of the society from 1887 to 1902 and taught sociology and the history of civilizations at the Ecole d'anthropologie from 1885. During the Empire of Napoleon III he had been only one step ahead of the repressive authorities as he contributed to ephemeral freethinking journals along with colleagues such as Hovelacque and the physician Henri Thulié. Letourneau was clearly committed to a sociological approach that allowed an important role for the milieu. He propounded an evolutionary scheme in which all races would eventually become civilized. Yet despite his overt socialism and lesser interest in physical measurements, he was also deeply implicated in theories of racial hierarchy and imperial discourse. For this reason Letourneau on racial (though not gender) issues represents nature masquerading as nurture.

The scientific materialists wished to expand anthropology to include sociology, linguistics, mythology, and demography. In 1882 Letourneau prepared a "questionnaire of sociology and ethnography" for the society that built on his work in Italy for the Florentine anthropologists in 1873. As Blanckaert argued, the categories based on nutritive, sensitive, affective, social, and intellectual needs stressed the organic and biological basis of all needs in line with an overriding materialist philosophy.[49]

Already in 1867 Letourneau postulated a three-stage narrative of development from societies expressing merely nutritive needs (Australian) to sensitive (African or Tahitian) or intellectual (Aryan) societies. Peoples in the early phases had larger occipital brain development, he claimed, while the affective societies had larger development of the parietal lobes.[50] The Lamarckian discriminatory paradigm then reared its head. In 1868 Letourneau, writing on the "psychology of blacks," thought it "insane to pretend to elevate rapidly the presently inferior races to the rank of the improved races." Only by a "gradual, systematic education acting in the same direction for numerous generations, and tending to develop their useful and superior aptitudes will they finish by being hereditarily transmitted."[51] Twelve years later he would still argue, "a race with inferior temperament has never created a superior

49 Blanckaert, "Le premesse dell'antropologia 'culturale' in Francia," 51, 56–7, 60–2, 67.

50 Letourneau, "Sur les phases sociales," 378–82; these brain-based distinctions came from Pierre Gratiolet's classification of racial development into occipital, parietal, and frontal lobe dominance; see Blanckaert, *De la Race à l'évolution*, 172.

51 Letourneau, "La psychologie noire," (1868) *Science et matérialisme*, 320–1.

civilization," since the "weight of the organic curse cannot lighten except with efforts of more than a millennium." "Milieu does much," he continued, "but it doesn't do all, and race is more important."[52] In his work on the evolution of property, Letourneau viewed colonization as tutelage by adults of children, not treating them as inferiors, but "owing them help and protection."[53] In one of his last major books, he hoped Europeans would become capable of truly humane conduct toward backward races as a "benevolent and patient tutor."[54] There was no hope of giving up this role as long as these backward races needed assistance along the evolutionary path to a future fraternal age. Letourneau's considerations on gender roles are discussed later in this chapter.

QUATREFAGES AND THE PARTISANS OF THE MILIEU

While Letourneau nominally sought to induce a sociological perspective that in reality masked an organic basis, some anthropologists harboured strong beliefs in the power of the milieu. In the early years, Armand de Quatrefages was a prestigious opponent of Broca's harsher hereditarianism. Without an equivalent resource network, Quatrefages gained respect as professor of anthropology (from 1855) at the Muséum d'histoire naturelle. He also wrote for elite intellectual journals of general interest such as the *Revue des deux mondes* and gave public lectures. During the American Civil War Quatrefages displayed his concern about American apologists for slavery by emphasizing common human nature in a "separate human kingdom," the only living creatures manifesting morality and religion.[55]

Like the naturalists Buffon and Etienne Geoffroy Saint-Hilaire, Quatrefages acknowledged limited variability in species, so that "race" was a "variety transmitted by generation." He was never reconciled to applying Darwin's ideas or any transformism to either animal or human evolution, and he also believed the limits of variation were stricter for animals than for men. Still he thought that, through migration and the action of climate over long periods, whites became bronzed in Ethiopia and Africans became lighter in Europe. Such changes were subsequently inherited. If, as the hereditarians argued, the races in Egypt on monuments look the

52 Letourneau, *La Sociologie d'après l'ethnographie*, 3–4, 26–7.

53 See Verrier, Review of Charles Letourneau, 93; for a somewhat different interpretation of Letourneau, see Blanckaert, "Les Prémices de l'anthropologie culturelle," *De la Race à l'évolution*, 461–97.

54 Letourneau, *La Psychologie ethnique* (1901), 98, 118, 128–30 (citation on 128), 244, 324.

55 Quatrefages, *Unité de l'espèce humaine*, 7, 20, 36, 47, 116–17.

same as the races today, the Egyptian milieu is responsible for subjecting them all to the same circumstances, not the separate creation of distinct races or species.[56]

Quatrefages thus believed that "both heredity and milieu work to produce races" through regional centres of creation, migration, and acclimation.[57] His protégé Franz Pruner-Bey also noted the importance of the milieu in producing races, since there was never a tropical white nor a temperate-zone native black.[58] Quatrefages maintained the belief that all races are not equally gifted, and conceded the literary, artistic, and architectural inferiority of Africans. But the milieu was more powerful for him than for many colleagues. Even the Australians could rise in a favorable milieu "above the condition in which we have found them, and would, in some respects, very quickly reach our level."[59] He earnestly urged his colleagues to "avoid denying to other races aptitudes, which remained latent for centuries in our ancestors before they were developed, and which are still in the same condition in too many of our fellows in civilization."[60] Hence where the ostensibly sociological Letourneau was a closet believer in organic supremacy on racial issues, Quatrefages, the partisan of unequal racial aptitudes, remained convinced that a favourable environment could reduce the physical significance of race for intellect and character.

Lesser-known society members also highlighted the influence of milieu. The statistician and demographer Louis-Adolphe Bertillon, who invented the term "mésologie" (study of the influences of the milieu), suggested extensive research to determine the relative influences of physical (climate and soil) and social milieu.[61] He hoped that the example of honest government could help change the social milieu. However, milieu was not all-powerful. Climate certainly differentiated Aryan peoples, he thought, but Australians or Chinese in ancient Greece or Italy could not have brought forth a philosopher like Aristotle or a poet like Virgil.[62] Bertillon still introduced a sociological orientation

56 Quatrefages, *Discours d'ouverture au cours d'anthropologie*, 5, 8–9, 11–12.

57 Quatrefages, *Rapport sur les progrès de l'anthropologie*, 129, 192, 314–17, 377–81, 386.

58 Muséum MS. 2258, letter to Quatrefages of Pruner-Bey, 1 mai 1860, 2 mars 1861, 28 mai 1861.

59 Quatrefages, *The Human Species*, 450–1, and repeated in the posthumous French edition of 1896, 332–3.

60 Quatrefages, *The Human Species*, 455; *L'Espèce humaine*, 336.

61 Bertillon, *BSAP* 4 (1863): 324, and "De l'Influence des milieux ou mésologie," 717–18, 720; see also "De l'Influence du milieu ou mésologie," 309–20.

62 "De l'Influence des milieux," 725–6 and "De l'influence des milieux sur nos idées et nos moeurs," 126–9, 131, 134; on Bertillon's conversion to a Lamarckian perspective, see Blanckaert, "Fortunes de Lamarck," *De la Race à l'évolution*, 421–3.

that most hereditarian colleagues avoided, and read his memoir to Emile Littré's Société de sociologie in May 1872 before reading it to the anthropologists in August.[63]

A weaker voice in the wilderness was the physician, philosopher, and experimenter in hypnotism, corresponding member Joseph-Pierre Durand de Gros (1826–1900), who pleaded for anthropologists to listen to hygienists as well as sociologists. Noting different physical conditions in different regions of the Aveyron, Durand thought that inadequate diet affected height, head shape, teeth, and even intelligence. He mockingly asked of strict hereditarians, "Why don't they see in the gaunt, puny, haggard population found in certain industrial cities that lives in attics or frequents houses of ill-repute, the descendants of a primitive race characterized by tuberculosis and skeletal deformations and why isn't this race condemned by the inexorable law of blood to eternal misery?" He hoped humanity could "struggle with success in making use of the forces of the surrounding world for its improvement."[64] For Durand, then, belief in the power of the milieu was a call to action.

Gustave Lagneau, an expert on the origin and diversity of ethnic groups in France, described the head and hair diversities among the Germanic long-headed and light-haired (Kimric), Celtic (round-headed, light or dark brown hair), and Aquitaine (long-headed, dark hair) groups. Still, he admitted that topography, diet, or social and professional role could influence their height or appearance. So could "vicious habits" or alcohol, which obstructed normal nutritive processes, and "excessive or anti-hygienic" occupations in mines or factories. Certain areas induced goitre or malaria because of mineral or climatic influences. In particular, Lagneau lamented the too often "antihygienic labour that industry and exploitation of mines motivate, the confinement in factories, the overly sedentary life of workshop and school, and the human encumbrance of our cities with narrow streets and multifloor buildings." This recognition of the significance of the urban and industrial landscape anticipated sociologists' concerns.[65]

In the generation after Broca, a strong voice for the influence of milieu was Jean-François, also known as Franz, Schrader (1840–1920), a

63 On Bertillon with respect to the more hereditarian Eugène Dally and Georges Hervé, see Blanckaert, "L'anthropologie lamarckienne," 620–4.

64 Durand de Gros, "Sur l'action des milieux géologiques dans l'Aveyron," 139–42; "Sur l'action modificatrice des milieux," 229; on Durand as an enthusiastic Lamarckian, evolutionary socialist, and adversary of Vacher de Lapouge, see Durand de Gros, *Questions de philosophie morale et sociale*, introduction by David Parodi, v-xxxvii, and 81, 96.

65 Lagneau, "Influence des milieux sur la race," 143–7 and *RMEA* 5 (1895): 130–3 (citation, 133).

cartography specialist for the publisher Hachette and anthropological geographer at the Ecole. Schrader argued, somewhat like Jared Diamond more recently, for Australian retardation because of dry climate, the lack of a great river valley, topography, and the lack of easily cultivated plants and easily domesticated animals. While Australians might be condemned to disappear, Africans, who may develop in an unconventional non-European way, should not be regarded as "instruments of work and gain." The "inferior race" of today could become the "superior race" of tomorrow. He was also one of the few to warn of the danger in destroying the environment by clear-cutting the forest, which he considered akin to using up the lives of workers in factories. Nor would destroying the Congo rain forest help the world rainfall supply. For him, as for other geographers like Paul Vidal de la Blache, there was reciprocal interaction between humanity and the natural world. In 1914, he showered praise on the American conservation movement, which represented the important insight that "Harmony between present humanity and the planet is indispensable to safeguard future humanity."[66] Schrader stands as one of the few anthropologists who were non-hereditarian, non-racist, and non-imperialist.

On the fringes of the anthropological establishment were Christian egalitarians such as the friend of Hamy, the marquis de Nadaillac. This anti-Darwinist Catholic conservative wrote that the inexorably harsh climate, famine, and the moral poison of whisky degraded the Eskimos. With confidence in missionary moralization, he hoped that perhaps civilized races that have become "savage and demoralized could one day "climb up the social ladder again."[67] Hence ideas of Christian brotherhood could also affect the inclination to monogenism and the perceived role of milieu.

THE MUSÉUM SCHOOL: HAMY, VERNEAU, AND DENIKER

The Muséum school of anthropologists, with some influence from Quatrefages on the next generation, would include his longtime collaborator, Muséum national d'histoire naturelle (MNHN) chair successor in 1892 and curator of the new Ethnography Museum of the Trocadéro in 1878, Ernest-Théodore Hamy (1842–1908), the next occupant in 1907 of the MNHN chair and Trocadéro director René Verneau (1852–1938), and the MNHN librarian Joseph Deniker (1852–1918).

66 Schrader, "Géographie anthropologique," 219–53, and (1897) 134, 138–40, and (1900): 122–5; on conservation, see "Les Lois de préservation aux Etats-Unis," 19.

67 Nadaillac, "Les Esquimaux," 94–104.

Given the occasional antagonism between the Société d'anthropologie de Paris or the Ecole d'anthropologie and the Muséum school, there has been an understandable tendency to stress the differences between them. Yet these differences, despite some bitter altercations, turn out to be less decisive than is apparent at first sight.

Hamy was caught between the social radicalism and uncompromising philosophy of the scientific materialists and the more measured Lamarckianism of Topinard, who wished to enlist him as an ally. This veritable virtuoso of networking among all potential colleagues was formerly Broca's lab assistant. From 1880 the Musée d'ethnographie became the repository for materials collected from travellers or agents in the colonies, underlining the importance of ethnography for its own sake.[68] Hamy's correspondence shows how Léon Rosny unsuccessfully attempted to ally with him by writing satires that would "caress the materialist group."[69] I will not discuss his *Revue d'ethnographie* (1882–89) here, except to say that it contained little theoretical reflection. The periodical offered an opportunity to discuss cultural artifacts, but also gave its readers a steady diet of anthropological measurements. Unlike the Société d'ethnographie, it did not try to keep its distance from the professors of the Ecole d'anthropologie. Hamy, like Quatrefages, believed that the milieu affected primitive division into races of a single human species, but that it had less power in more recent times.[70]

To some degree, Hamy's successor René Verneau advocated a stronger role for the milieu and integrated the ethnographic approach with the anatomical. As Williams has noted, Verneau in his youth conducted an extended expedition on natural and artificial rock formations on the Canary Islands, but was so frustrated with his poor pay from the Muséum that he almost gave up anthropology to engage in provincial medical practice.[71] In 1892 he delivered lectures on anthropology for the Université populaire de la Ville de Paris. In 1894 he acquired the important post of editor of *L'Anthropologie*, where until 1930 he would display some hospitality to new trends and to colonial officials seeking opportunities for publication. In 1902 he became president of the Société d'anthropologie.

68 A.N. AJ 15 663, Dossier de Hamy; see Dias, *Le Musée d'ethnographie* for an overview of Hamy's importance.

69 Muséum MS. 2312, Letter of Rosny to Hamy 18 juin 1888.

70 Hamy, "La terre et l'homme," *Revue de géographie* 18 (1886): 15–26, 81–90, cited in Dias, Le Musée, 231.

71 Muséum MS. 2257, Letter of Verneau to Hamy, 11 septembre 1889 [actually 5 September 1890] where a "Touraine medical practice" seems more lucrative than anthropology; cf. Williams in "Anthropological Institutions," 343.

His early work rejected the British theorist Robert Knox's idea that races could not adapt to other climates. Verneau, like Quatrefages, thought African-Americans had already reduced by one-quarter their alleged gap in development from whites, and he also favoured emancipation of women, though he expected them voluntarily to remain housekeepers and mothers. In 1910 his anthropology course continued to argue that, since extant races have obviously already changed, the milieu could modify both "customs and the skeleton, muscular systems, even the brain."[72]

However, the very order of presentation of the races in his first book was tendentious, being arranged according to level of civilization. An egregious passage on the Jews, classified as Semites, degenerated into a diatribe against a group who "think they own the whole earth and have the right to take back by ruse what was stolen from them."[73] Verneau was not receptive to studies that argued the milieu could change the cephalic index, including Franz Boas's contention that the American environment altered (but was not completely responsible for) the cephalic indices of Sicilian and eastern European immigrants to the United States. He thought Boas's conclusions "quite improbable, given what ethnic anthropology has shown in the most diverse regions."[74] In a popular lecture delivered in 1899 during the Boer War, in addition to denouncing war as a savage practice, Verneau argued for limits to the power of the milieu. The Boers, in the same physical environment as the Hottentots and Bushmen, were certainly not becoming like them.[75] Moreover, at the Neuchâtel International Congress of Ethnology in 1914, Verneau adamantly opposed Arnold van Gennep's proposal to protect ethnography

72 Verneau, "Discours d'ouverture, cours d'anthropologie," 268.

73 Verneau, *Les Races humaines*, 8–9, 31–2, 38, 188, 557–8 on Jews, 651 on women; see Deniker's uncomplimentary "Review of Verneau," which criticizes his lack of scholarly apparatus and "fantastic" illustrations, 386.

74 Verneau, Review of Adolphe Bloch, 547–8; *Institut français d'anthropologie*, 21; cf. Léon Poutrin's critique of Boas's statistics or of any statistics to solve anthropological problems in *L'Anthropologie* 24 (1913): 309; transformation of cephalic index had already been debated at the 1900 Congress of Anthropology and Prehistoric Archaeology; see *L'Anthropologie* 12 (1901): 547–8 where Verneau criticizes Bloch's claims in "De l'Origine des brachycéphales," 541–7, by saying that the milieu cannot operate in two different directions if it has influence; see also Silva Telles, "La Dégenerescence des races humaines," 241–3, where he argues that the Portuguese degenerate in an African environment; for Bloch, see also "De la transformation d'une race," 73–83, and "Sur une prétendue découverte," 207–8, on Boas's inflated claims of originality; for a recent critical appraisal of Boas, see Sparks and Jantz, "A reassessment of human cranial plasticity," 14, 636–9; and the response by Gravlee, Bernard, and Leonard, "Heredity, Environment, and Cranial Form," 125–38 and (2003): 326–32.

75 Verneau, "Les Boers et les races de l'Afrique australe," 903.

from physical anthropology. Van Gennep had wanted to "separate sections of ethnography from physical anthropology since they harm one another." Verneau thought this "scattering of forces" would work against the goal of "co-ordinating conclusions."[76]

In 1910, however, the heir of scientific materialism Adrien de Mortillet believed that the Muséum faction in the society was contesting the work of the commission studying racial mixing. Verneau resigned from the society in this dispute and Mortillet was involved in a fistfight with the psychologist Henri Piéron, a "liegeman of Verneau." More resignations from the Muséum group followed.[77]

In the interwar period, an intermediate position on nature and nurture remained characteristic of Verneau, rather than an alleged Muséum school stress on nurture alone. He equated Australians with Neanderthals, and saw that "most backward races have evolved within certain limits."[78] But he now admitted that the colonial administrators Louis Faidherbe and Maurice Delafosse had convinced him that in view of early African metallurgy and animal domestication, the "Negro can progress on his own."[79] So even the later Muséum Lamarckians, while conceding some progress for all, would not necessarily envisage development to the European level and were less optimistic than Quatrefages.

The other Muséum representative, the librarian Joseph Deniker, an immigrant from Russian central Asia with French parents, became president of the Société d'anthropologie in 1904 and a member of both van Gennep's Institut ethnographique international (1910) and the sociologically oriented Institut français d'anthropologie (1911). In his Broca lecture to the Société d'anthropologie in 1898, he assumed that one "cannot deny the influence of race on uncultivated peoples," but among more civilized peoples, "social causes have more importance than physical causes for individual psychic development." Thus differences between Australians and Fuegians may be explained on racial grounds, but among developed societies, the occupation of individuals – priests, bankers, or miners – may promote common character traits. One day one might "appreciate the respective value of different types who may become a part of the general progress."[80]

76 Verneau, "Congrès international d'ethnologie et d'ethnographie," 367, 370 on apolitical anthropology.

77 Wartelle, "La Société d'anthropologie," 160.

78 Verneau, *Les Origines de l'humanité*, 55–6, 61–2, 73.

79 Verneau, *L'Homme, Races et Coutumes*, 5, 11 on transmission of acquired characteristics, 273–5 on Jews, 386–7 on Africans.

80 Deniker, "Les Races et les peuples d'Europe," 198, 200, 206; Deniker's essay commissioned by Alfred Binet, "La Question des races en psychologie," 306 (citation).

Two years later he asserted that whether a "somatological unit" (a group with "physical characteristics combined in certain ways") was called race or species mattered little, so he declined to adjudicate the monogenist-polygenist dispute. Like Topinard, he classified groups by skin colour, hair type, and nasal or eye shape into twenty-nine secondary and six main divisions.[81] Language, religion, and social institutions, on the contrary, constituted "ethnic groups," a term he popularized, and ethnography was therefore descriptive.[82] Deniker also testified to diminishing returns obtained from anthropometry. Unlike anthropologists of Broca's era, he thought no anthropometric measures consistently correlated with intellectual ability. Deniker was one of the orthodox anthropologists thus responsible for the crisis in the anthropometry he practiced so well.[83] He concluded that there was "no scientific demonstration of the psychic forces related to each race," and included remarks on the artistic talents of Australians and Bushmen and the rich mythology of Polynesians.[84]

In the posthumous second edition of his larger survey, he continued to argue that milieu could change the chemistry, but not the morphology, of bodies. Therefore, Boas's statistics were doubtful.[85] He remained undecided on the most crucial question of whether race or milieu was more influential, since cranial capacity and brain weight were only weakly correlated with intelligence, but milieu could work more easily on the civilized. Climate alone did not change skin colour. By now he seemed lukewarm on European colonization, since the cost of adapting to other climates might be too high. In time, southern Europeans, crossed with other races, might live in the tropics.[86] Once again, the receptivity to milieu was a matter of both innate endowment and civilization. Not all peoples could profit equally from the benefits of milieu, but those living with the advantages of economic and cultural refinements could more reliably display further improvements.

MANOUVRIER: THE APPARENT TRIUMPH OF MILIEU

The career of the last major anthropologist discussed here, a onetime disciple of the Broca school, Léonce Manouvrier (1850–1926), illustrated

81 Deniker, *Les Races et les peuples de la terre* (1900), 2, 4, 8–9, 11.

82 Ibid., 37–9.

83 Ibid., 82, 90, 118.

84 Ibid., 129–30, 148, 466; see Verneau's lukewarm Review of Deniker, 762–6, citing mistakes on Australian marriage and expressing doubts about racial divisions.

85 Deniker, *Les Races et les peuples de la terre* (1926), 46–7, 73, 93, 137–8.

86 Ibid., 141–2.

two phenomena about turn-of-the-century anthropology: the countercurrent contesting abuses of craniometry and openness to consideration of "social facts." Along with Quatrefages, the Muséum school, and the aforementioned members of the Société d'anthropologie, Manouvrier did not believe physical nature all-important. His record on gender issues, however, is mixed.

Manouvrier, as an anatomically skilled physician, worked in Broca's anthropology lab at the Ecole pratique des hautes études from 1880, and taught physiological anthropology, general anthropology, and ethnology at the Ecole d'anthropologie from 1883. He directed the lab at the Ecole pratique from 1903, and succeeded Letourneau in 1902 as general secretary of the Société d'anthropologie. He also worked for the natural history lab of the Collège de France, and delivered lectures as an alternate for the history of science chair in 1910 and 1912. His interests were diverse: he was president of the new Société de psychologie in 1904 and one of three vice-presidents of René Worms's Institut international de sociologie in 1910; he also taught a course at the Ecole d'anthropologie about the relationships among the three social sciences in 1909–10.

Manouvrier gained fame for his opposition to Cesare Lombroso's anatomical markers for the criminal. He was also a fellow traveller with the Gabriel de Mortillet circle of scientific materialists, who published the periodical *L'Homme* in 1884–87.[87] However, in later years he was a bitter opponent of the ambitions of Gabriel's son Adrien de Mortillet to occupy space and gain influence in the anthropology lab.[88] He also conducted a scarcely concealed war against hard-line hereditarians such as Georges Papillault.[89]

In his Broca lecture of 1890 on "aptitudes and acts" Manouvrier warned his colleagues against creating a new phrenology in which individual aptitudes were expected logically to produce certain actions. People now realized, he argued, that the emergence of individual modifications by natural or sexual selection depends on the milieu. Inborn aptitudes limited neither the capacity of women to reason nor the ability of "Negro" children in the United States to become as literate as whites.

87 For Manouvrier's career, see Archives nationales F17 23860, Dossier de Léonce-Pierre Manouvrier; for criminal anthropology, see Mucchielli, "Hérédité et 'Milieu social'," 189–214; Blanckaert, "Des Sauvages en pays civilisé," 55–88. Hecht, "A Vigilant Anthropology," 221–40, and *The End of the Soul*, 211–56; Blanckaert, "La crise de l'anthropométrie," 95–172; Topinard, Review of Lombroso, 475–8, also clearly opposes Lombroso's markers for criminals; on the scientific materialist circle, see especially Richard, "La Revue *L'Homme* de Gabriel de Mortillet," 221–56.

88 Archives nationales, F17 23860, Memo to [Jean-Baptiste-Vincent] Laborde, 1901.

89 For Papillault's views on morphology and intelligence, see Raoul de La Grasserie, "La Race et le milieu," 387–90.

African students in France learned algebra and literary criticism "with the same aptitudes of their ancestors in savage life." Therefore, the inherited elementary "aptitudes propose, but the milieu disposes."[90]

While Manouvrier never minimized the significance of brain size, other things being equal, he admitted that smaller brains might have similar or greater energy than larger brains, and that, given the proportions of body mass, "the mental inferiority of women disappears before a profound psychological analysis." He continued to promote the importance of education, family, social milieu, and occupation.[91] Manouvrier also went on to develop theories of psychic energy and temperament that required other measures than brain size.[92] Always an advocate for both the biological and sociological approaches to anthropology, he argued that "human societies act to such an extent on individuals composing them that one can hardly have an idea of what the thoughts, sentiments, and acts would be of a man who, since childhood, would have been completely deprived of all social influence."[93]

As a member of René Worms's Institut international de sociologie in 1895, Manouvrier wished to react against exaggerated emphasis on race and heredity.[94] During the furor over the Dreyfus affair in 1899, Manouvrier wrote a mordant attack on the hereditarianism of Le Bon and Lapouge. Jennifer Hecht has found evidence that the socialist deputy Gustave Rouanet requested the article and reviewed it favourably in his newspaper, *La Petite République*.[95] If people thought they could not escape the "infirmity of race," as Le Bon argued, they would not make the maximum effort at self-improvement. Indeed, no one could adapt if all abilities were inherited. Even in Lamarckian terms, a father's efforts could only transmit "improvement of elementary aptitudes," not habits.[96] Contrary to Lapouge's

90 Manouvrier, "Les Aptitudes et les actes," 941–3, 949–50; for his plea to consider social causes of crime rather than "apelike" or atavistic traits of criminals, see "L'Atavisme et le crime," 228, 230, 233–5.

91 Manouvrier, "Essai sur les qualités intellectuelles," 70, 72, 76, 80, 84 (citation).

92 Blanckaert, "La Crise de l'anthropométrie," 133–5; see also Bloch, "L'Intelligence est-elle en rapport avec le volume du cerveau?" 577–619, for questioning of whether brain size alone, independent of impressionability or excitability of cells, is sufficient for intelligence.

93 Manouvrier, "L'Anthropologie et le droit," 366.

94 Manouvrier, Discussion of Golberg, "L'Origine des races et la division du travail," 370, 372–4, 379–80, 383, 385; cf. Paligot, *La République raciale*, 54–6; Mucchielli, *La Découverte du social*, 145–50; see discussion of this intervention in chapter 7.

95 Manouvrier, "L'Indice céphalique et la pseudo-sociologie," 233–59, 280–296; Hecht, "A Vigilant Anthropology," 232–4, 239.

96 Manouvrier, "L'Indice céphalique," 237, 239.

new "craniomancy" or "new phrenology," the mixed populations of France had no reason to hate each other because of cephalic index, or hair or eye colour. Assuming Aryans were blonde, would all blondes be superior in intelligence, character, and morality?[97] There was "no relation between the determinations of intelligence and moral character and morphological variations of the cranium expressed by the cephalic index," which varied even among individuals of the same people. Contrary to Lapouge's expectations of dolichocephalic dominance, Balzac, Laplace, and Pascal were all brachycephalic.[98] "In spite of the ingenious efforts of its promoters, this system is to be rejected from anthropology and sociology; it's only pseudo-science."[99] The Durkheimian Henri Hubert favourably reviewed this article in the 1901 *Année sociologique.*[100]

While the passions unleashed by the Dreyfus affair may be partly responsible for Manouvrier's attack on the cephalic index, the Ecole d'anthropologie gave at least two other signs that it was not at this time a vehicle for anti-Semitism. The professors reprinted a speech by the pro-Dreyfusard founder of the League of Rights of Man, Francis de Pressensé, on the need to make France the most respected nation in the world. The editors also published a translation of the materialist and socialist Carl Vogt's attack in 1880 on what he perceived as shameful German anti-Semitism.[101] However, the labelling of Jews as a race did not disappear entirely among the anthropologists. Aside from René Verneau's diatribes, the Jewish deputy director of the Ecole d'anthropologie Henri Weisgerber criticized Boas's assistant Maurice Fishberg for concluding that the Jews were not a race. In 1911 he maintained that the author was mistaken because he was acquainted only with American Jews.[102]

However, given Manouvrier's lukewarm reception of the Haitian lawyer Anténor Firmin's work on the equality of Africans, one cannot conclude that he had discarded concepts of racial hierarchy.[103] When he presented his candidacy to the Collège de France in 1899, he continued to speak of the importance of anthropology in explaining not only individual differences but differences between the "currently most inferior races" and the "most civilized." Differences between races and sexes

97 Ibid., 244, 248, 251, 255.

98 Ibid., 280, 294.

99 Ibid., 296.

100 Mucchielli, La Découverte, 288–9; Zerilli, *Il lato oscuro,* 37.

101 *RMEA*, 9 (1899): 95, 153–9; Mortillet had opposed anti-Semitism in "Anti-Sémitisme," 522–8, though he thought that without persecution Jews would assimilate; on the other hand, he thought, in "L'Anti-Sémitisme" (1887): 606–7, that German Jews were oversensitive about a society dedicated to repealing their emancipation.

102 [Weisgerber], Review of Fishberg, 407.

103 Manouvrier, Review of Anténor Firmin, 180–2.

might have, he continued, a "more or less direct relation with intellectual and moral character."[104]

Yet at the same time, when writing not as an ingratiating applicant for a position but as a scholar, Manouvrier had not repented by 1911 his previous denigration of anthropometry. His essay on "Anthropometry and Aptitudes" for the Brussels International Congress on Pedology (study of children) argued that faith in such measures for classification and education of children was often misplaced. Although he had helped the psychologist Alfred Binet in head measurement, he now criticized "fanatical calculators for the value of aptitudes," a reproof of considerable import considering Binet's earlier curiosity about such measurements.[105]

Sports champions could not be predicted from proportions of body limbs. Rather with "cultivation and training, the average subject could become distinguished in all sports."[106] Trials of athletes or candidates for employments would be better predictors than measurements. Even brain volume, despite its importance, could be applied profitably only 2 or 3 per cent of the time to judge individual cases. Cerebral mass was not an indicator of memory, ability to work rapidly and assimilate information, presence of mind, or linguistic skill. Simple visual inspection would enable the elimination of extremely unsuitable individuals, and otherwise "physiognomy would be about as reliable" as measurements.[107] In a seeming slap at the deceased Binet, he concluded, "The measures in the mental order are in effect comparable with respect to the insufficiency of physiological and psychological analysis that they represent, to the anatomical measures aimed at above, and they ensue from analogous illusions to those that I have wanted to place in evidence."[108] Manouvrier remained therefore a harsh critic of quantitative evaluation that was not firmly grounded on true correlations of the physical with mind and character. While not giving up his residual belief that brain size and composition matter, he introduced a strong sociological perspective into anthropology that, despite inconsistencies on gender, was more coherent than the sociology of Letourneau.

The most satirical critique of anthropometry came from the Belgian physician Gaston Daniel, who wrote on Congolese and other Africans. All anthropological columns of figures seemed to him "columns of indices whose meaning escapes everyone, even those who invented

104 Archives du Collège de France, G-IV-f 45f 1897–1900, "Considérations présentées par M. L. Manouvrier à l'appui de sa candidature, 7 nov. 1899."

105 Manouvrier, "Anthropométrie et aptitudes," 413.

106 Ibid., 424–5.

107 Ibid., 477–8, 481–3.

108 Ibid., 487.

them."[109] He noted sardonically that the alienist Désiré Magloire Bourneville formerly published the dimensions of male genital organs of the "young idiots" of Bicêtre. Why not, he added, did he not continue with the diameter of the anus, the depth of the vagina, and the color of feces? While he thought blacks and whites "of a different essence," Daniel also believed blacks could evolve in a few years through stages that took centuries for Europeans.[110]

ANTHROPOLOGISTS ON GENDER

Gustave Le Bon represented one extreme of the argument that the intellectual and moral capacities of women were sexually determined and based partly on general volume of the cranium. He had even maintained in 1879 that with increasing civilization the gap between men and women would increase. There was no hope that education would enable women to catch up.[111] Paul Broca had clearly argued on the other side that education would increase brain size, including for European women. Broca himself may very well have encouraged reservations in the society review of Le Bon's essay. Charles Letourneau acknowledged that civilized men might have increasingly larger brains because of the "increasing difficulty and requirement of intelligence of masculine occupations that the social state makes more complicated, while women hardly leave family life and receive a more or less limited education." The solution, however, was not to conclude that women were incapable of closing the gap. Rather women deserved a "virile instruction and education, so faculties would not gradually atrophy. Le Bon arrives at a completely different conclusion. For him, woman is seemingly an accursed being, like poor Ham in the Bible, and he predicts abomination and desolation if women leave home, which seemingly makes her more brutish. We naturally have all kinds of reservations about this conclusion." Letourneau even went so far as to point out how brain volume alone was not enough. It is "no easy task to appreciate the relation between cranial volume and degree of intelligence," he admitted. In the brain "quality can often compensate for quantity of nervous substance" since "many kinds of intelligence exist and there is no common measure for them."[112] Yet Le Bon's impact was so powerful that no less an anti-hereditarian sociologist

109 Daniel, "Essai anthropologique sur les nègres," 357.

110 Ibid., 371, 376.

111 Le Bon, "Recherches anatomiques et mathématiques," 101–4.

112 Letourneau, Review of Le Bon, "Recherches anatomiques et mathématiques," 382–3; Letourneau, Review of Le Bon, "L'Homme et la société," 332–6, citation on 336; on Topinard and Le Bon, see Le Bon, "Correspondance," 747–8, and Topinard, "Correspondance," 748; see also Wartelle on Broca's reservations about Le Bon in "La Société d'anthropologie," 136.

than Emile Durkheim in his early work on *Division of Labour* admitted that comparing "the volume of the skull of a man or a woman ... presents considerable differences in favour of the man."[113]

Moreover, in a group renowned for openness to improved status of women, Letourneau sounded at first only marginally more sympathetic than Gustave Le Bon. Only a year after this report (1880) Letourneau thought that to advance, women would have to "masculinize their brains by suitable instruction," since their condition was a fact of anthropology "beyond any controversy." After many generations, they could contemplate political equality.[114] This description was certainly more open-minded than the physician Henri Thulié's "mystic" or "courtesan" alternative for women in his articles a bit later in *L'Homme*, where he denied women suitability for abstract thought. Thulié opposed making women doctors or engineers. He believed women have their own special qualities and should not struggle competitively with men.[115] He was one of the few anthropologists who held equally intolerant views on matters of race and gender.

Still, by the time of his death Letourneau was much more flexible about the capabilities of women. He now argued against an essential organic nature of women that created a "feminine type." Rather, social conditions in civilized societies resulted in a division of labour that deprived women of intellectual culture and insulated them from an active and free life. In the future he hoped for full equality for women in all spheres so their aptitudes could be freely developed.[116]

Letourneau thus deserves a reputation as a supporter of feminism and seems more open-minded on women than on "inferior races." The case of Manouvrier is far more complex than usually acknowledged. Manouvrier had attended the first International Congress on the Rights of Women in 1889. He also showed that, given the ratio of brain weight to body mass, women's smaller body mass and proportionally smaller brains were no signs of intellectual inferiority. Even more important, the degree of intelligence of anyone "depends on many external conditions that are independent of the organic."[117] Yet despite his belief in the

113 Sowerwine, "Women's Brain, Men's Brain," 290.

114 Letourneau, *La Sociologie*, 181.

115 Henri Thulié, "La Femme, Education normale," 256–61 and "De la Femme, Essai de sociologie physiologique," 229–46.

116 Letourneau, *La Condition de la Femme*, 501–7.

117 Mahoudeau, Review of Léonce Manouvrier, "Sur l'Interprétation de la quantité de l'encéphale," 318–9; Fee, "Nineteenth-Century Craniology," 415–33.

equality of women's intelligence, he retained notions about women's destiny that made him a half-hearted feminist.[118]

In an 1883 ethnology lecture he postulated that anthropology could help determine the "best fashion of distributing labour among the two sexes generally envisaged." He suggested that women as soldiers and men caring for children had not yet been proven appropriate social roles. But there was a "need to reform general economic and social conditions" imposed on women since men were too egotistical in denigrating capacities of women. Implicitly this was an attack on Le Bon's viewpoint and explicitly on the anti-feminist legal expert Léon Giraud, who attributed the inequities of the Civil Code to a divine curse. Manouvrier found no valid disabilities in the feminine organism as such. Eugène Dally had already said that the differences with women were exaggerated in civilization, and the roles of culture and nature were difficult to determine, since women were soldiers in Dahomey.[119]

Manouvrier insisted that anthropological science supported the "amelioration of the fate of woman and the elevation of her social condition, which could have the happiest consequences for the interest of the child, of man himself, and for society."[120] In a series of lectures on the applications of the anthropology of the sexes in 1909, Manouvrier reminded his audience that he had concluded women's brains were larger in relative weight nearly thirty years earlier.[121] However, he still believed that it was not the lack of education, but the severe selection and competition of men in the higher professions that made it difficult for women to catch up. Previously he had supported the rights of women to be witnesses to legal acts and to control their own earnings, and of women medical students to be admitted as interns in Paris hospitals. None of these rights would threaten the "separation of the domains of the sexes."[122]

Manouvrier's ultimate position about women's role in society was only marginally different from someone such as Topinard. In a phrase reminiscent of eighteenth-century physicians, he remained convinced that "sexual differentiation penetrates the totality of the organism."[123] Women's need for protection during pregnancy, nursing, and early child rearing mandated not only the feminine affective character, but also the traditional sexual division of labour. A "bio-sociological truth" therefore required maternal and familial roles complementary to the male provider

118 On the early papers, Hecht, "A Vigilant Anthropology," 224–6.
119 Manouvrier, "Cours d'ethnologie," 1–2, 4; Dally, "Femmes," 427–39.
120 Manouvrier, "L'anthropologie et le droit," 360.
121 Manouvrier, "Conclusions générales sur l'anthropologie des sexes," 41, 44.
122 Ibid., 43, 54, 57.
123 Ibid., 44–5.

and protector.[124] The mother would have to "strive to assure to her husband the happiness that he sought mutually, as the wife did, in conjugal solidarity." In addition, "to propose the confusion of social functions and intersexual competition is a new symptom of the evil emerging from similar sociological absurdities." A reasonable feminism would therefore work to open more occupations to women, but not compete with men or oppose women's nature. Otherwise women would remain childless or neglect their children and their households.[125]

While Manouvrier's position somewhat resembled the attitudes of contemporary sociologists on necessary social roles, the biology of an essential feminine nature seemed to underlie these declarations. They cast doubt on the definitive victory of nurture over nature that seemed implicit in Manouvrier's other formulations. The qualifications were also much more representative of the anthropological mainstream on gender issues, while Letourneau was exceptionally progressive toward the end of his life and Broca had set the pattern for somewhat modest encouragement of women's educational gains.

Both Letourneau and Manouvrier encouraged the first woman admitted to the society, Clémence Royer. Her militant feminism (though she was not a suffragist) was in a way opposed to the whole thrust of her evolutionary philosophy that sanctioned the outcome of the process. For she thought that the use of reason and intelligence in modern society could assure the education of women and their ability to overcome hereditary instincts of submissiveness and the selection for beauty and passivity that had so far been the norm.[126] Women were saved, she said in the preface to Darwin, by the strength they inherited from male ancestors. But now the despotism of husbands and male dominance was an obsolete situation contingent on ancient circumstances. She hoped for equal rights of spouses before the law. She also saw the primitive sexual instinct as beneficial in view of social customs that made motherhood painful and shameful.

No doubt the feistiest statement on the role of women ever articulated in the Société d'anthropologie was her 1874 comment giving a woman's perspective on the low birth rate in France.[127] She launched a blistering indictment of men for whom woman is the "animal in all creation about

124 Ibid., 46–50, esp. 49.

125 Ibid., 50–1, 61 (long citation).

126 Harvey, "*Almost a Man of Genius,*" 144–5 on moral philosophy, 162 on suffrage.

127 "De la natalité," MS, Archives de la Société d'anthropologie de Paris reprinted in Harvey as "Clémence Royer on Women, Society, and the Birthrate," "*Almost a Man of Genius,*" 193–203; discussion on 125–32.

which man knows least."[128] She attacked bourgeois marital customs in which marriage is a "profession" for women, the lack of opportunity for women to be educated, to support themselves, to fulfill properly the role of motherhood, and to have the right to divorce. In fact, the role of motherhood, comparable in her view to military service, was in part to furnish sons to seize land from "inferior" races. Royer represents a refutation of the theory of the equivalence of all the marginalized, since she wished to liberate women, but not "dominated" non-Europeans. To minimize legal disabilities to a portion of the population, illegitimate births should be treated equally to legitimate births. She attacked the emotional insensitivity of young men and the Church for its idealization of celibacy and marital restraint, which she thought morally equivalent to the use of abortion by rich American women. In line with her harsh social Darwinism, she also thought that celibacy might be imposed or abortion and infanticide might be permitted in France for physically or mentally incapable parents.[129]

When she would not consent to editorial changes in her article, the Société d'anthropologie de Paris decided to suppress its publication. This may not have been in itself explicitly anti-feminist, though Broca certainly disagreed with her views sanctioning harsh natural selection. Delicate negotiations with the medical faculty, the Paris municipal council, and prospective financial backers were underway at the time for the establishment of the Ecole d'anthropologie. The national government of "Moral Order" might also have looked askance at such radical questioning of social customs.[130] However, the censorship testified to the difficulty for a full and fair hearing of feminist arguments even in a society where several significant members were sympathetic.

THE DISCOURSE OF IMPERIALISM: POLYGENIST FEARS AND MONOGENIST OPTIMISM

The anthropologists largely subscribed to the conventional wisdom that Europeans deserved to rule over other peoples who were either permanently or temporarily subordinate in the racial hierarchy. However, there was no monolithic imperial discourse partly because the most militant believers in racial superiority were polygenists who at first questioned the feasibility of Europeans' ability to adapt to other, especially

128 Harvey, "*Almost a Man of Genius,*" 195.

129 Blanckaert, "L'anthropologie au féminin," 34–6; Harvey, "*Almost a Man of Genius,*" 58 on women in military service, 66 on selection for passivity.

130 Harvey, "*Almost a Man of Genius,*" 133.

tropical, climates.[131] Paradoxically, the more tolerant monogenists, who believed in the superiority of the milieu, were sometimes more enthusiastic for empire because of their confidence in acclimation. Certainly almost all anthropologists hoped their findings would be useful for colonial policy – either for or against – and they willingly participated in the much-visited Colonial Expositions of 1878, 1889, 1900, and 1931. Gradually the majority became more reconciled to a form of imperial rule that preserved local customs but guaranteed French domination.

In the early years of the society, concern about high mortality of soldiers and settlers in Algeria made the debate about acclimation more urgent.[132] For example, the Marseille physician Jean-Christian-Marc Boudin opposed the idea that humans could flourish anywhere and, even before the society's foundation, Georges Pouchet, had written about the suitability of each of several human species to its geographic domain.[133] However, one convenient solution to the colonial dilemma would be to find races more suitable to colonial climates, such as the "Iberians" of southern France for Algeria. Settlers and soldiers could then intermarry with indigenous peoples to form a better-adapted race that would guarantee French rule.[134]

French military misadventures in Indochina in 1885 not only overthrew the ministry of Jules Ferry but also provoked continuing misgivings among anti-imperialist anthropologists. Gabriel Mortillet, though he believed in the present cultural and morphological inferiority of Africans, opposed imperialism because of the dangers of the climate. As a socialist deputy in 1885 he questioned the "murderous colonial policy" of sending French soldiers to the "murderous climate" of Tonkin.[135] The republican journalist, anti-socialist, and free trade economist, patron of the anthropologists in the Paris City Council, and also deputy for Paris and future Public Works Minister Yves Guyot produced a classic document published in the same year, *Lettres sur la politique coloniale.* Guyot also believed that racial crossing was the only solution to European adaptation since Europeans survive with difficulty in torrid climates.[136] Like economic opponents of empire, he lamented the costs of supporting

131 Blanckaert, ed. *Les Politiques,* 19, 98.

132 Blanckaert, "La Crise," 113–4.

133 Ibid., 119–20.

134 Ibid., 123; Paligot on Bertillon, *La République raciale,* 61–2.

135 Mortillet, *L'Homme* 2 (1885): 701; Blanckaert, "La Crise," 114–24, esp. 116 on Quatrefages, 118–19 on Périer; Osborne, *Nature, the Exotic, and the Science of French Colonialism,* 62–97.

136 Guyot, *Lettres sur la politique coloniale,* 17–28; see also Paligot, *La République raciale,* 63–4, 93.

colonial warfare to gain customers with little purchasing power when free trade among developed nations might bring equivalent commercial benefits and investments could be better employed at home. He also repeatedly warned that the empire was a distraction from the defence of French security in Europe.[137]

Most remarkably Guyot, in the midst of the furor over Indochina, formulated one of the most comprehensive attacks on the "civilizing mission," echoing Georges Clemenceau's critiques of empire. Relying on his own journeys to north Africa and incidents from the *Bulletin de la Société de protection des indigènes*, he railed against European ferocity, arbitrary taxation and requisitioning, and misunderstanding of native property in Algeria.[138] He also questioned racial hierarchy in a passage alleging that Fuegians are less different from Parisians than one Parisian is from another in a different milieu. The visible improvements in Maori, Canaks (New Caledonians), Australians, and African-Americans showed the educability of these races. He mocked the idea of expansion of a definable French race since there was no pure French race.[139] He suggested that, rather than killing other peoples, the French could bring them knowledge and justice through education and medical assistance. In an unusual inversion of imperial logic, he argued that Europeans could learn from Hindu, Arab, and Chinese civilizations, and from the communal institutions of the Kabyles.[140] The mission to civilize, as contemporaries interpreted it, masked rapacious instincts and was incompatible with a republican government founded on universal male suffrage.[141]

So both humanitarianism and belief in traditional racial hierarchy could result in anti-imperial pronouncements. True enough, Guyot was not really an anthropologist himself, but an interested amateur who had associated with anthropologists in freethinking circles and who later became director of the Ecole during World War I. But even the mainstream anthropologists still raised questions, though from a rather more uncomplimentary view of non-Europeans. The linguist André Lefèvre was a staunch hereditarian who considered Africans so uncivilizable that they could never become truly enlightened. In 1892 he asked "would it not be much wiser to leave them to themselves?"[142] His friend Abel Hovelacque however thought that better fertility and life expectancy of civilized peoples burdened them with a surplus population so

137 Guyot, *Lettres sur la politique coloniale*, 99–105, 113–14, 213, 335, 344.
138 Ibid., 156–7,159, 177–9.
139 Ibid., 216, 357–8, 396, 422–4.
140 Ibid., 277, 355.
141 Ibid., 368, 394.
142 André Lefèvre, "Races, peuples, langues de l'Afrique," 65–79, citation on 69.

colonization was inevitable. Inferior races would "exhaust themselves and disappear."[143] Europeans should peaceably offer guidance, but the "black was a big child on whom no hopes can be founded … European civilization cannot meet his needs." While colonization was therefore justified, Hovelacque wondered if for humanitarian reasons the African should be spared brandy, missionaries, and rifle shots.[144]

In 1961 Raymond Betts noted that Le Bon had been the forerunner of imperial "associationist" views at colonial congresses that saw no possibility of rapidly educating inferior races. Le Bon remarked, "Negroes' brain development is in the Stone Age" so that it would take centuries to "transform the mental abilities of Africans." With such inequalities, "inferior races cannot surpass a certain level," and colonial civil servants would be well advised to understand that natives would never achieve political rights, though it was worth studying their language and customs.[145] Le Bon clearly influenced the formulation of "associationism" by retired naval officer Léopold de Saussure and other speakers at the Colonial Sociology Congress of 1900.[146] Yet even Le Bon sometimes lapsed into pessimism. In 1911 because of the "impenetrability of races," he wrote, like Guyot, that colonization could be renounced without great loss because there were so many unexploited races within France.[147]

By contrast, some monogenists who were most sympathetic to the capabilities of other races also recommended pursuing the imperial mission. Quatrefages suggested that after a period of higher morbidity and mortality, Europeans could acclimate to other zones since migrations were part of the history of the human species.[148] Even the gentle Hamy, as Nélia Dias has already pointed out, in his Sunday lecture series at the Muséum, admitted that he had been " for long years an ardent defender of our colonial expansion." He continued, "My devotion to this great cause has perhaps increased more when circumstances placed me in more intimate contact with the region of the Niger," where he realized that "soldiers, missionaries, and explorers are up to the admirable task of pacification and progress assigned them, and I rejoice patriotically about this."[149]

143 Hovelacque and Hervé, *Précis*, 192; *L'Homme* 2 (1885).

144 Desmet, "Abel Hovelacque," 77–8.

145 Betts, *Assimilation and Association*, 59–89; Le Bon, "Ethnographie, Congrès international pour l'étude des questions coloniales," 227–8, 234–5.

146 Sibeud, *Une Science impériale pour l'Afrique?*, 69.

147 Duprat, Review of Le Bon, "La Psychologie politique et la défense sociale," 327.

148 Blanckaert, "La Crise," 115–18; Paligot, *La République raciale*, 59–60.

149 Hamy, "Tombouctou," in *Revue générale des sciences pures et appliqués* 13 (1902): 972; cited also in Dias, *Le Musée*, 233; see also his eagerness to co-operate with the Muséum on investigation of tropical conditions in a memo in AN AJ 15 556.

In later years as mortality in Algeria declined, the anthropologists became more likely to support colonial ventures. Colonial governors either before or after their tenure in the administration also joined the society. Paligot has commented on the participation of the physiologist Paul Bert, later resident-general in Indochina (1886), who wanted a civilizing mission and hoped the native would assimilate or disappear. Similarly, naval physician Jean-Marie-Antoine Lanessan later became military governor of Indochina (1891–4) and navy minister (1899–1902). Lanessan defended republican colonization as the equivalent of migrations in ancient times, opposed brutality, but clearly wanted Vietnamese as "associates" in a protectorate. Explorer Louis Faidherbe also continued his linguistic and geographic studies in part with the anthropologists after his return from the governorship of Senegal[150]

ANTHROPOLOGY FOR "SCIENTIFIC" COLONIALISM: "ASSOCIATIONISM"

Would-be colonizers, even if they did not themselves measure heads, paid attention to the anatomical studies and theories of racial hierarchy of the anthropology society, including their implications for intellect and character. Naval and military physicians trained at the medical faculty near the headquarters of the society. As corresponding members, they travelled the world to examine head shapes and later used Anders Retzius's cephalic index comparing the diameter of the head front-to-back with the width as a ratio multiplied by 100. They also examined facial features for resemblances to Europeans. Chief naval physician Laurent-Jean-Baptiste Bérenger-Féraud (1832–1900) commented on the potential of the Wolofs of Senegal as a work force substituting for Europeans since they were superior to other Africans from the viewpoint of prognathic facial features[151] Ten years later, Dr Armand Corre (1841–1908), a tropical medicine expert and colonial criminologist, still argued that the "transformation of blacks is fatally limited by their particular cerebral organization." Since Africans were unassimilable to European civilization, France should develop Senegal and Algeria while southern European white settlers could develop the rest of Africa.[152] Whatever their reflections on the environmental influences on susceptibility to disease that would mitigate the indelible

150 Paligot, *La République raciale*, 66–7, 69–72, 74–5.
151 Bérenger-Féraud, "Etude sur les Ouolofs (Sénégambie)," 469, 484, 494; see, however, Tremeau de Rochebrune's critique in "Etude morphologique, physiologique, et ethnographique," 294, 298.
152 Corre, "De l'Acclimatement dans la race noire africaine," 93, 96–7.

imprint of race, there was no surrender of a hierarchical attitude that could justify empire.[153]

In 1875 Jules Assézat outlined the "scientific" future of colonization in a review of the debate. He summarized the then standard assumptions – Europeans from warm climates such as Italy would do better in Algeria and crossbreeding and creation of an intermediary race, even with all its risks of moral decline, would be the best way of acclimation.[154] Anthropologists of the second generation, such as the medical geographer Arthur Bordier, aspired to a "scientific" colonialism that combined racial hierarchy with ideally peaceful methods and adjustments to local temperaments. He made no concessions on the supposedly retarded brain development of Africans and chemical differences in the blood that meant that "the Negro cannot be civilized except by force."[155] Nor was racial crossing rapid enough in promoting cerebral improvement. Bordier never surrendered the goal of French rule. He warned that African Muslim action against France was "in vain" and that France should re-take Tonkin, especially since the Annamese "natives appreciate our domination."[156]

But in principle he still opposed war and pillage, and advocated "pacific association" that would produce mixed race offspring.[157] As with "associationist" principles the "indigenous civilization would be directed, not crushed by us." He envisaged independence some time in the future and a time when worthy peoples like Kabyles would send their "co-religionists to the Chamber and Senate" in France.

Similarly, the deputy from Oran, Camille Sabatier, portrayed Algerian policy as a "question of anthropology." The French could win over the Berbers, but should give up trying to civilize Arabs.[158] "Anthropological evidence" showed "not all groups have an equal part in indefinite perfectibility" since "physical, intellectual, and moral aptitudes are hereditary."[159] Unlike Bordier, he foresaw nothing beyond participation in communal

153 Fogarty and Osborne, "Constructions and Functions of Race," 206–36, use later revisions in Bérenger-Féraud to argue that pathology was not seen as indelibly racial, but as a product also of environmental and social influences.

154 Assézat, "L'Acclimatement," 294–305.

155 Bordier, "L'Hérédité. Mécanismes et Théories," 4 (1894): 327–8; "Le Milieu intérieur et l'acclimatation," 1 (1891): 140–42; Paligot, *La République raciale*, 57–8.

156 Bordier, *La Colonisation scientifique et les colonies françaises*, 289, 419, 498.

157 Ibid., xiii, 16, 42–5, 47, 498; Martinet, in Review of Bordier, 330–1, agrees on the goals to educate, civilize, and promote racial crossing.

158 Sabatier, "Les Berbères sédentaires," 442, and "Etude sur la femme Kabyle," 56–69; Lorcin, *Imperial Identities*, 159–63.

159 Sabatier, "Du Progrès indéfini de l'humanité," 326, 334, and "De l'Evolution du progrès dans les groupes sociaux," 417, 429, on Kabyles.

assemblies for Kabyles. In the last analysis, citadels and the use of force would overcome insurrection.[160] The same author foresaw a twentieth-century war of Europeans against the menace of the Chinese and the yellow race.[161]

In lectures on the "Hovas" (Merinas) of Madagascar, Sigismond Zaborowski, an ethnography professor, made his course an account of "ethnology of French colonies" as an adjunct to empire, though he was among the few anthropologists who questioned imperial ideology. Still, he would later oppose immigration of Africans which he felt threatened the "Aryanism" of France.[162] Zaborowski, in an exchange with the Tunis military physician Lucien Bertholon in 1897, defended a much more benevolent imperial policy than many of the Algerian settlers, despite a rigid separation of the races. In offering physician Alfred Etiévant's brochure to the society, he doubted the possibility of assimilating Muslims, but thought they should be allowed to retain their differences in customs and practices as "loyal associates" of the French enterprise. In addition, the study of anthropology showed the "extreme diversity of the character of peoples" so that one could "learn to be just to each of them, to sympathize with their needs and their legitimate aspirations."[163] The veterinary expert André Sanson and physician René Collignon agreed with his sentiments.

Bertholon furiously replied with ample out-of-context translations from the Koran that Muslims, deadly adversaries of Christians, were not assimilable. He would rather increase the non-French European population in north Africa for a labour supply while urging Muslims to accept new land concessions in the Sudan. France could thus profit from racial divisions among Sudan peoples and also ensure dominance of Algeria. Zaborowski insisted that he was only condemning certain kinds of settler behaviour rather than disapproving of the entire enterprise of colonization.[164] In the ensuing discussion, Sanson did not think "any enlightened or wise mind would mean by assimilation the granting of all civil and political rights."[165] But France's domination of Algeria needed to become accepted by the native population through benevolent and just

160 Sabatier, "De l'Art du peuple éducateur," 38, 41–2, 44.

161 Sabatier, "Essai sur l'ethnologie de l'Afrique du nord," 459.

162 See the opening lecture in his 1921 course cited in Sibeud, *Une science impériale*, 270.

163 Zaborowski, Review of Etiévant, "Choses de l'Algérie," 140; see also Paligot, *La République raciale*, 57.

164 Zaborowski, "De l'Assimilation des indigènes algériens," 490–92; Bertholon, "Quel doit être le rôle de la France," 509–36; reply by Zaborowski and discussion by Sanson, 587–604.

165 Ibid., 601–2.

treatment and sound institutions, rather than the prevailing scorn of soldiers and settlers.

With regard to Morocco, Zaborowski was one of the frankest opponents of hypocrisy in the civilizing mission. He asserted that it would suffice to intervene in Morocco for the French to say that they needed to defend Algeria, the Sahara, and Africa, regardless of any mission of saving Moroccans from barbarism. In short, strategic objectives were sufficient to justify expansion.[166] He continued, "We have no right to change violently a people's customs and much less to attack its beliefs." Most interventions did not in fact result in general improvement of the material and moral existence of a people, as witness the passive resistance of Turks, Muslims, and Moroccans. So he understood perfectly well the ideological function of slogans. Algerian chieftains thought their existence as good and beautiful as ours. Nothing in reason, justice, or law would convince them otherwise. "It would be a shame to treat peoples under our tutelage with hateful feelings, symptoms of a total incomprehension." So even among the racists the evolutionary schemes could seem wanting as justifications of conquest. Zaborowksi saw through all the claims of European moral superiority made by some of his colleagues.

The medical geographer Franz Schrader devoted his 1912 lectures to Morocco. Earlier he had been a strong advocate of the influence of the milieu and had even questioned the meaningfulness of the term "race." Schrader argued that assimilation should not mean domination for some, coupled with effacement and subjection for others. He pleaded for sympathy with the Islamic state of mind. He, too, thought Berbers, unlike Arabs, could evolve since they created the Moorish civilization in Spain. France should free them from their present decadence by taking their hand, and initiating them into a higher ideal.[167]

Colonial administrators did not, however, give up the civilizing mission and were not averse to using scholarly justifications of racial hierarchy that made imperial rule appear benevolent or inevitable. Paligot has demonstrated the attention paid to the pro-imperial anthropologists' pronouncements by Jules Harmand, who had definite ideas about the moral superiority of Europeans in Indochina. The explorer and official Edouard Foa in Dahomey took head measurements and concluded that Africans had thieving and lazy temperaments. The Congo governor Adolphe Cureau in central Africa wrote of the fatalistic and melancholic attitude of Africans – a psychology allied to a racial temperament. Few

166 Zaborowski, "Les Méditerranéens et les nègres, la dépopulation et les colonies, les Marocains et L'Europe," 149–92.

167 Schrader, "L'Afrique du nord maghrébin," *Revue anthropologique* 23 (1913): 76–88.

were distant from theorists of racial heredity. The political lobbyists of the Union coloniale such as Joseph Chailley-Bert (son-in-law of Paul Bert) expressed similar interest in theories that justified limiting the rights and duties of races with limited aptitudes.[168]

"Scientific" colonization aspired to send colonists who could survive and adapt better. Members of both the Société d'ethnographie and the Société d'anthropologie recommended respect for existing cultures that were deemed worthy of progress. While some members of both groups disapproved of violent conquest, neither group consistently suggested that European intervention was not needed for civilizing purposes. At best, authors such as Zaborowski thought that expansion could be justified by French strategic needs rather than by pretending to benefit other peoples. He agreed, though, like Bordier that the methods needed to be pacific and as mild as possible under the circumstances.

CONCLUSION

Among the orthodox anthropologists, at the outset Broca and Quatrefages had disagreed on the relative importance of milieu in the establishment of races, but almost all of the anthropologists of the next generation were Lamarckians who admitted that milieu was of considerable importance. The major qualification was the differential Lamarckianism that classified Europeans as adaptable while most others were not. Topinard, for example, believed Europeans could inherit acquired moral habits, while other peoples remained relatively unchangeable. Letourneau granted an evolutionary trajectory for all peoples, but suggested that race still wins out over milieu, and European guidance was necessary for playing appropriate roles in the historical script. A few hard-line hereditarians, such as the notorious Le Bon and Vacher de Lapouge, outside the mainstream of anthropologists, believed the milieu was largely powerless. They were either political conservatives or radicals of the right who held out little hope for progress, predicted inevitable racial struggle, and feared the decadence of the most promising Europeans.

The so-called Muséum school largely assumed the heritage of Quatrefages that foregrounded milieu. Yet differential Lamarckianism still triumphed. Verneau came around to allowing some progress to all peoples, though not necessarily to the same level. He remained adamant on the fixed cephalic index and the indelible nature of the Jews. Deniker demolished the usefulness of indices of anthropometry, and never quite admitted a scientific basis for psychological assumptions about races.

168 Paligot, *La République raciale*, 221–2, 227, 230–2, 235.

However, he remained convinced that milieu could assert its influence more easily on the civilized. Only Manouvrier remained the most consistently inclined to admit other environmental influences as most important not only in the formation of criminals but also in transforming all aptitudes into actions. Like Deniker, he realized that there was a crisis in anthropometry. Yet while he thought brain size an insufficient index, he remained somewhat ambivalent on the potential of all races to progress.

More remarkably, after denying that the smaller size of women's brains made them intellectually inferior, and agreeing with Broca that exercise of women's brains could have remarkable effects, Manouvrier remained an essentialist on the social role of women. In theory, the more support for hereditarianism, the more an anthropologist should have insisted on limiting women's capacities. But the case of Manouvrier illustrates that essentialism could rear its head even among anthropologists perceived as friendly to feminism. While Le Bon was the most relentless in denigrating women's capacities, Letourneau near the end of his life went far beyond Manouvrier in advocating equality for women in the family and society. In the debate about women's role in society, the anthropologists would be much more reticent on the whole than the sociologists. They allowed Royer to voice her feminism at meetings, but refused to publish her most sensational and provocative arguments.

Anthropologists' theories of racial hierarchy were more rigid if they were polygenists who separated humanity into distinct species. But monogenism was not a sufficient condition for advocacy of equality. In imperial discourse the roles at the outset were somewhat reversed. The rigid believers in racial hierarchy who questioned the influence of the milieu doubted the adaptability of Europeans to other climates. Thus they would not vociferously support a French imperial mission. By contrast, monogenists assumed that migrations had always influenced the adaptation of human varieties; hence European acclimation was feasible, if at some cost. A few anthropologists genuinely sympathized with subject peoples. Guyot spoke of wasted resources, and Zaborowski doubted that the French benefited the peoples of the empire. However, even evolutionists such as Letourneau, who believed all peoples would progress, thought it would take so long to catch up that Europeans would have to act as tutors guiding others to progress. The most common trend among anthropologists was to give up on the ideal of assimilating all peoples as French citizens, but to seek less brutal means to "associate" peoples with French objectives and, if possible, to retain indigenous institutions. Few were like Schrader, who would deny the right of the French to pursue colonial rule either for a civilizing mission or for strategic and economic objectives.

Anthropology thus gave nominally scientific prestige to cultural assumptions about European superiority. Paligot has pointed out that the right to rule trumped all rights of subject peoples. Objections to European rule were usually couched in terms of feasibility rather than desirability. The crisis in anthropometry, however, and the inability to agree on standards for defining race, along with the rise of challenges from newly organized groups of sociologists, would lessen the predominance of physical anthropology.

While sociological approaches began to affect the emphasis of anthropologists, psychologists began struggling to assert their autonomy from philosophy and to strive for prestige as a science. A characteristic manifestation of this trend was to investigate the importance of heredity in human intelligence and character.

4

Ribot and Psychological Heredity

THOUGH THERE WAS OBVIOUSLY no single anthropological perspective, the mainstream heirs of Broca had postulated a hierarchy of physically defined racial groups correlating with a hierarchy of intelligence and character. These anthropological assumptions infiltrated the discourse of the new science of experimental psychology. Debate swirled around the issues of inherited national character and indelible individual personality.

RIBOT AND THE AUTONOMY OF PSYCHOLOGY

In late nineteenth-century France, all the human sciences struggled for an institutional place in the sun and for conceptual legitimacy that justified their emancipation from older traditions of philosophy or medicine. Psychology was still a subtopic of the philosophy curriculum.[1] Disciples of the "Eclectic" philosopher Victor Cousin still held sway in the Faculty of Letters at the Sorbonne. Their entire teaching was based on such concepts as an immortal spiritual soul autonomous from the body and the unity of the individual self in combatting destructive or anti-social impulses. By the 1850s Auguste Comte and other positivists who favoured the natural sciences as the paradigmatic route to pursuing knowledge had challenged the fiefdom of the Eclectics.[2] The foremost challenger was the quasi-positivist Hippolyte Taine, who had difficulty in obtaining his post at the Ecole des Beaux-Arts because of his suspected materialism. Taine had linked the biology of the nervous system to psychology and argued for literary criticism informed by the influences of

1 Brooks, *The Eclectic Legacy*; and "Philosophy and Psychology at the Sorbonne, 1885–1913"; Fabiani, *Les Philosophes de la République*; Nicolas, *Histoire de la psychologie*; Logue, *From Philosophy to Sociology;* Mucchielli, *La Découverte du social.*

2 On the eclectics, Goldstein, *The Post-Revolutionary Self.*

heredity, environment, and history.[3] In June 1873 the defence of a Paris Faculty of Letters thesis on psychological heredity by philosophy teacher Théodule Ribot (1839–1916) was a great event, perceived by reviewers in the press as a challenge to the old philosophical order. The members of his examination jury, Cousinian spiritualist philosophers Paul Janet and Elme Caro, were willing to pass Ribot's thesis, though they worried about its potential for aid and comfort to materialists.[4]

Ribot, who already had a mildly anti-clerical reputation as a lycée teacher, had earlier in his Introduction to *English Psychology* (1870) attempted to argue that metaphysics and extra-scientific speculation had no place in psychology and that introspection would never be sufficient to plumb the depths of mind.[5] Though not an experimentalist himself, he investigated the pathology of mental attributes or conditions – attention, will, personality – since these pathologies were the perceived equivalents of nature's experiments on humanity. Three years after his thesis examination, Ribot would found the *Revue philosophique*, a periodical open to all philosophical schools, but designed to give full scope to naturalistic psychology. While John Brooks suggests continuing Eclectic influence on Ribot's philosophy and Mara Bertolini points out that spiritualists were never shut out of the *Revue*, Laurent Mucchielli has established that Ribot intended his network of Ecole normale graduates, physicians, and other collaborators to shatter the old Cousinian consensus.[6]

Until recently little attention was paid to Ribot, but now the studies of Serge Nicolas and those of the research team of Jacqueline Carroy and Régine Plas have opened this period of French psychology to substantive investigation.[7] For years before the fateful thesis defence, Ribot had conducted a campaign to liberate psychology from its constraining metaphysical presuppositions with his fascination for English psychology, for

3 On Taine, Kahn, *Science and Aesthetic Judgment;* Charlton, *Positivist Thought in France*; Weinstein, *Hippolyte Taine*; Nordmann, *Taine et la critique scientifique.*

4 Ribot, *L'Hérédité, étude psychologique.* Two major articles focus on the role of heredity in Ribot: Bénichou, "Ribot et L'Hérédité psychologique" and Nicolas, "L'Hérédité psychologique d'après Théodule Ribot"; on Ribot, see also Nicolas and Murray, "Le Fondateur de la psychologie 'scientifique' française'"; Nicolas, Marchal, and Isel, "La psychologie au XIXe siècle"; Nicolas, "L'introduction de l'enseignement de la psychologie scientifique en France; Marchal et Nicolas, "L'emancipation de la psychologie"; Nicolas and Charvillat, "Introducing Psychology as an Academic Discipline in France"; Nicolas, *Théodule Ribot, Philosophe breton.*

5 For the school inspectors' attempted refutation of Ribot's reputation for materialism, see Dossier de Ribot, A.N. F 17 21608; *La Psychologie anglaise contemporaine*, 6–8.

6 Thirard, "La fondation de la 'Revue philosophique"; Bertolini, *Il pensiero e la memoria;* Mucchielli, "Aux Origines de la psychologie universitaire en France."

7 Carroy and Plas, *Histoire de la psychologie en France XIXe-XXe siècles.*

the evolutionism of Herbert Spencer, whom he translated, and the work of Charles Darwin.

Psychology was to "claim its autonomy from the metaphysical yoke" by avoiding unresolvable questions such as the nature of the soul, while it would "constitute itself as an independent science."[8] Ribot would be even more peremptory about the vanity of metaphysics in the introduction to *La Psychologie allemande contemporaine* (1879), which was enthusiastically reviewed by the physiologist and member of the Société d'anthropologie, Charles Richet.[9]

Despite early doubts about the focus and range of anthropology, Ribot's evolutionism provides a fine example of the permeation of anthropological assumptions concerning collective psychology, though he would later fitfully emerge from the anthropological shadow. In January 1880 Paul Broca and Paul Topinard presented him as a member of the Société d'anthropologie, though he never assumed a very active role.[10] After examining his views and the opinions in the periodical he founded, we shall also look at another great pioneer of experimental and educational psychology, Alfred Binet (1857–1911).[11] Binet represents the continued fascination with physical anthropology methods even though he realized that anthropometry would not provide him with satisfactory measures of intelligence. The two men were linked in that Ribot, as lecturer in psychology at the Sorbonne and Collège de France, obtained for Henry Beaunis the first experimental psychology lab at the Sorbonne. Binet then volunteered to join Beaunis in the lab in 1891 and in 1894 became its director. Binet would move from an early concern with the psychology of reasoning and mental functioning in hypnosis and in other laboratory experiments to study alterations of personality and ultimately to do research on individual psychology and child behaviour. Rather than evolutionism, it was his fascination with measurement that attracted Binet to anthropological techniques.

THE LEGACY OF TAINE AND SPENCER

The greatest critic of the Eclectics in this era was the historian of literature and would-be psychologist, Hippolyte Taine. Early in his career Ribot read Taine's work, corresponded with him after Taine had reviewed his thesis

8 Ribot, *English Psychology*, 4, 11, 18–19; *La Psychologie anglaise*, 6, 14–15, 22–3.

9 Introduction to *Psychologie allemande* republished in Marchal and Nicolas, "L'émancipation."

10 Archives de la Société d'anthropologie, Musée de la science de l'homme, C1 2335.

11 Wolf, *Alfred Binet*; Avanzini, *Alfred Binet et la pédagogie scientifique*; Andrieu, *Alfred Binet, sa vie, son oeuvre.*

favourably, and met colleagues at Taine's salon. In his 1863 introduction to English literature Taine had first asserted that there are very marked differences of moral and mental organization among peoples not understood by eighteenth-century philosophers. Each people had a "very general disposition of mind and soul, either innate and attached naturally to the race, or acquired and produced by some circumstance applied to the race." The three different sources were "race, milieu, and moment." [12] Differences in bodily structure and temperament revealed hereditary dispositions, though climate and society were also significant. Taine had appropriated the anthropological assumption that each physically defined group had correlated psychological characteristics, though environmental influences might inflect them.[13] Southern Europe, for example, had more sociable and polished peoples, while despotic governments elicited religions of charity and love, variable by the historical moment.[14]

Ribot also found in the philosophy of Herbert Spencer a full-blown naturalistic, Lamarckian evolutionism as influential in France as Darwin's theory. Ribot and his friend Alfred Espinas translated the first edition of Spencer's *Principles of Psychology* (1855) in 1870, only to leave it unpublished until Spencer authorized translation of the second edition of 1870–72 (1873–4 in French). For Spencer, the development from savagery to civilization illustrated the general principle of evolutionary change from simple and homogeneous to complex and heterogeneous. The distinction between biological and cultural heredity became blurred in adaptations inherited as modifications in the nervous system and those with a range gradually affecting the entire society.[15] "Primitives" might be able to adapt to a new milieu, but some inferior races manifested arrested development. Races could be warrior-like or pacific, nomadic or maritime, independent or slave-like. The long legacy from previous generations, including racial dispositions, was psychic heredity as well as physical heredity – memory becoming a transmissible instinct. Across the generations varied dispositions related to habits of life gradually strengthened, and the "tendencies to certain combinations of psychic changes became organic."[16]

Like Taine, Ribot consequently warned in the introduction to *La psychologie anglaise contemporaine* that psychology had too often studied

12 Taine, *Histoire de la littérature anglaise*, 1: xxii-xxiii.

13 Ibid., xxi-xxvii.

14 Ibid., xxix-xxx.

15 Gissis, "Herbert Spencer's two editions of the Principles of Psychology."

16 Spencer, *Principes de psychologie*, 1: 448–50; On Spencer's influence in France, see Mucchielli and Becquemont, *Le cas Spencer*.

white, male, civilized humans, "without worrying about black or yellow inferior races, as if human faculties were all identical in nature and vary only in degree, as if difference in degree cannot often be such as to become difference in kind."[17] This reminder of differential development was hardly a plea for tolerance of diversity. Carl Vogt, a member of the Société d'anthropologie, also provided Ribot with arguments about the differences in brain size among races and between savages and the civilized.[18]

Ribot's introduction to *La Psychologie anglaise contemporaine* highlighted human physical differences, such as the larger and more massive legs of Europeans compared to Papuans.[19] Even at this early stage he repeated without comment Spencer's affirmation that "inferior races are not thinking creatures" since their numerical concepts barely go beyond the counting capacities of a dog.[20] Similarly, he cited emotional differences, since "savages ignore justice and pity" and "love of property" is a late development. Ribot noted that Spencer had criticized the psychologist Alexander Bain for ignoring "hereditary transmission" that "creates the great difference between the savage and civilized races."[21] Another possible source for Ribot's interest in collective psychology was the German physiologist Ewald Hering, who in 1870 interpreted "organic memory" as an inherited property of organized matter.[22] So the germ of the thesis of hereditary psychology was already outlined in *English Psychology* (1870). Anthropological concepts mattered because Ribot relentlessly stressed the consequence of physical heredity for the psyche.

HEREDITARY DISEASE

French physicians who discussed the transmissibility of disease fostered the study of heredity as a phenomenon.[23] Only gradually was the term used for physiological heredity. Since the operation of heredity in the organism was hardly contested, and since the organism influenced psychological phenomena, it made sense to Ribot to complete the syllogism – psychological phenomena were also inherited. No doubt the impetus

17 Ribot, *English Psychology*, 17; *La Psychologie anglaise*, 25; Carroy and Plas, "La Méthode pathologique," esp. 607.

18 Ribot, *English Psychology*, 132, 159, 171; *La Psychologie anglaise*, 172, 203–4.

19 Ribot, *English Psychology*, 132.

20 Ibid., 159.

21 Ibid., 228.

22 Gasser, "La Notion de mémoire organique dans l'oeuvre de T. Ribot," esp. 309–10; Otis, *Organic Memory*, ix.

23 Lopéz-Beltrán, "Forging Heredity"; "In the Cradle of Heredity."

behind Ribot's interest in heredity increased beginning in 1873 when he audited medical courses and began his friendship with the neurologist and psychiatrist Jean-Martin Charcot, director of the lab at the Salpêtrière hospital in Paris and famous analyst of hysteria.

A physician member of the Phrenology Society (though not a phrenologist himself), Prosper Lucas (1805–85), formulated the first serious studies of heredity in a thorough, prolix two-volume treatise in 1847–50. Perhaps most importantly Lucas introduced Ribot to the idea that heredity and (pre-Darwinian) variation were both manifestations of the same natural force. For Lucas in his fixist, anti-transformist framework, aside from perpetual resemblance, there was also an "innéité" (innateness) producing difference. In practice "invention and imitation were two laws of nature."[24] Such a view enabled Ribot to characterize heredity as a universal force since either one or the other aspect of the force could explain a great range of phenomena. He also absorbed from Lucas the strong conviction that nervous aptitudes and diseases are inherited.[25]

Few authors in any era minimized the influence of age, sex, social milieu, and profession. But Lucas continually returned to the "native im pulses of organization, as well as the first roots of the passional tendencies of the individual" so that "le naturel" (inborn nature) was the principal source of human actions and of nervous aptitudes. Education could awaken or stimulate mental faculties, but not originate them.[26] So heredity was a predisposing cause particularly noticeable in criminals and the mentally disturbed, who had natural inclinations to wrongdoing rather than learning them from their families. Lucas saved free will by arguing that impulses could be controlled and that morality was a matter of refusing temptation.[27] Like Ribot later, he was suspicious of efforts to intervene to create talented races, since he thought the principle of variation would work against them. Lucas also thought that "idiocy and imbecility" were inherited and that national character was proven to be transmissible by ethnological facts.[28] At the same time he thought that climate and diet could change human races (even in head shape).[29]

Just a few years later an oft-cited work by the alienist Jacques Moreau de Tours (1804–1884), an expert on mental illness and experimenter with hashish, appeared.[30] While he was teaching in the Vésoul lycée from 1866

24 Lucas, *Traité philosophique et physiologique de l'hérédité naturelle*, 1:24.
25 Ibid., 1: 235.
26 Ibid., 1: 527, 545, 566.
27 Ibid., 1: 455, 457, 486, 494.
28 Ibid., 1: 445, 577, 586.
29 Ibid., 2: 310, 465.
30 Jacques Moreau de Tours, *La Psychologie morbide.*

to 1868, Ribot read Moreau de Tours.[31] Like Lucas, Moreau de Tours asserted that education could not create the native disposition of the faculties. More reluctant to use physical indicators, he still thought a physical predisposition in organs of thought – a real lesion – created madness or idiocy.[32] He also believed in hereditary inclinations to drunkenness and crime. Most tellingly, he believed 90 per cent of mental illness hereditary, though precipitated by alcohol or belladonna (nightshade extract used as an anti-spasmodic).[33]

One of the most influential treatises on pathological heredity was the work of Bénédict-Augustin Morel, the inventor of degeneracy theory.[34] Degeneracy was a morbid deviation from a primitive human type – a series of "vicious organic dispositions over several previous generations" that could transform and worsen in each successive generation. Eventually it might be associated with an incurable fatal condition, but along the way idiocy, sterility, and feeble-mindedness were possible.[35] These vicious dispositions could sometimes be judged by external appearances of poorly conformed heads or other deformities and sometimes by aberrant behavior. Moral degenerates, such as instinctive torturers of the weak, could never improve if released from asylums.[36] To Morel, however, it was clear that many environmental or social factors could provoke mental disturbance or criminality. The "law of double fertilization" entailed environmental influences acting on hereditary dispositions. A bad education could bring out direct predispositions to alcoholism. Moreover, poor soil, a swampy countryside, poor diet, or unhealthful industrial occupations could provoke degeneracy. Morel was generally confident that moral treatment was not futile for many disorders, but once true degeneracy had begun, there was no turning back. Heredity therefore accelerated disorders initiated by the milieu.[37]

GALTON AND DYNASTIES OF INTELLIGENCE

With the evolutionary approach of Spencer incorporating organic memory and the alienists' conviction of the inheritance of pathological

31 Ribot, *Psychologie anglaise*, 398–9, and Nicolas, "Hérédité psychologique," 308.

32 Moreau de Tours, *Psychologie morbide*, 9, 19.

33 Ibid., 111–2, 116–7.

34 Morel, *Traité des dégénérescences physiques, intellectuelles et morales de l'espèce humaine*; Coffin, "Heredity, Milieu and Sin"; Coffin, *La Transmission de la folie, 1850–1914*; Pick, *Faces of Degeneration*; Nye, *Crime, Madness, and Politics in Modern France*.

35 Morel, *Traité*, 1: 61–2, 565.

36 Ibid., 1: 118.

37 Ibid., 1: 113; Coffin, "Heredity, Milieu, and Sin," 157–63.

conditions, all Ribot needed now was a plethora of case studies showing the inheritance of intelligence in families. The evidence came from Darwin's cousin Francis Galton (1822–1911), a polymath who from his rather unsuccessful university days was fascinated by the reality of intellectual inequality. After his geographic and ethnographic expeditions to southwest Africa in 1850–52, he reflected on the inequality of human "races" and then outlined in two papers for *Macmillan's Magazine* in 1864–65 the speculation that would energize his entire intellectual career. In "Hereditary Character and Talent" he asserted that "I can only say that the general resemblances in mental qualities between parents and offspring, in man and brute, are every whit as near as the resemblance of their physical features; and I must leave the existence of actual laws in the former case to be a matter of inference from the analogy of the latter."[38] Whole families of talented people were more common, Galton thought, than lone outstanding individuals.

Using biographical dictionaries of eminent persons and minimizing social and educational influences, Galton initiated the line of reasoning that he would later call eugenics: "The general intellectual capacity of our leaders requires to be raised, and also to be differentiated. We want abler commanders, statesmen, thinkers, inventors, and artists. The natural qualifications of our race are no greater than they used to be in semi-barbarous times, though the conditions amid which we are born are vastly more complex than of old. The foremost minds of the present day seem to stagger and halt under an intellectual load too heavy for their powers."[39] Therefore, one needed to encourage marriages in the elite castes, and retard them in the lower-ranked intelligences. There has been much controversy over whether Galton had deeper underlying ideological motivations for his interest in heredity or whether he was merely seeking membership in the Darwinian clique by expressing interest in the transmission of traits.[40]

The work of Lucas and George Lewes led Galton to believe that character is also inherited. This would include "a craving for drink, or for gambling, strong sexual passion, a proclivity to pauperism, to crimes of violence, and to crimes of fraud." Similarly, Galton was convinced that races differed in character and intellect. Native Americans were "naturally cold, melancholic, patient, and taciturn" through all the different climates and political jurisdictions of America. Yet they all had "local attachments" and "an astonishing sense of personal dignity." Africans

38 Galton, "Hereditary Character and Talent," 158.

39 Ibid., 166.

40 Cowan, "Nature and Nurture"; Forrest, *Francis Galton, The Life and Work of a Victorian Genius*; Waller, "Putting Method First"; Gillham, *A Life of Sir Francis Galton.*

had "strong impulsive passions, and neither patience, reticence, nor dignity." Each was "endowed with such constitutional vigour, and so prolific, that his race is irrepressible." Even altruism and religious sentiments stemmed from native instincts explicable as a part of character.[41] From the outset in the preface to *Hereditary Genius* (1869) he asserted that the "modern European possesses [natural ability] in a much greater average share than men of lower races." His own experiences in southwest Africa had led him to such a conclusion.[42] Therefore inferior races needed to be guided by European civilization or else supplanted.[43]

While Ribot was not interested in statistics or in Galton's opposition to inheritance of acquired characteristics, other aspects of Galton's work were important to him. In *Hereditary Genius* Galton again outlined his opposition to natural equality, refuted in the nursery, school, university, and professions.[44] Nevertheless, Galton said he never denied "the great power of education and social influences in developing the active powers of the mind." The point was that a maximum of intellectual performance potential existed just as there was a maximum of muscular strength for a blacksmith.[45]

Galton's refinement of the theory for English scientists led to the famous proclamation, "Nature is all that a man bring with himself into the world; nurture is every influence from without that affects him after birth."[46] In responding to the botanist Alphonse de Candolle's critiques, he still argued that within the same race and country, nature was far stronger than nurture, though education was needed for genius to manifest itself. However, he was now ready to concede that intelligence was not so strongly hereditary as more elementary qualities of mind and body.[47] But in the next year when, like modern psychologists, he sensed the importance of twin studies, he felt he could not escape the conclusion that "nature prevails enormously over nurture when the differences of nurture do not exceed what is commonly to be found among persons of the same rank of society and in the same country."[48]

41 Galton, "Hereditary Character," 320–3.

42 Galton, *Hereditary Genius*, x.

43 Galton, "Hereditary Character," 332.

44 Galton, *Hereditary Genius.*

45 Ibid., 12–13, 327–8.

46 Galton, *English Men of Science*, 12.

47 Ibid., 195.

48 Galton's twin studies were published in "Short notes on heredity" and "The history of twins"; Forrest, 132.

Alphonse de Candolle wrote the most direct reply to Galton in 1873. Ribot could not have known about this work for his first edition, although his examiner Elme Caro raised it during Ribot's thesis defence. The chief quarrel of Candolle was Galton's refusal to acknowledge that "traditions, examples, and inner family councils have more influence than heredity, properly speaking."[49] In his own survey of scientists and foreign associates in state academies, Candolle hailed the importance of temperate climates, small countries, relatively high living standards, and absence of religious dogma and of hindrances to inquiry in education. Even Candolle admitted, however, an inheritance of general ability and mathematical skill. For Candolle, education, example, reflection, and an infinity of social causes modified heredity. [50]

Most important from the perspective of Ribot, Candolle conceded to Galton the importance of race as a factor. For Candolle "all European nations have more or less contributed to the advancement of science, while the Asian, African, and indigenous American races have remained, to the contrary, completely outside the scientific movement." The Chinese and Japanese could not rival European science, and among the great human races, it is evident that the "white race is more intellectual than coloured races." Mental and moral flaws of other races were the significant differentiating factors.[51] Galton felt that Candolle had substantiated his own position in admitting the significant importance of race, and Ribot had no reason to think otherwise, even when he became cognizant of Candolle's position. His review of the second edition of Candolle in 1885 criticized him for not looking at the pathology and transmission of mental illness. But he approved of Candolle's agreement that heredity was a general law though it might be impossible to know what was inherited and what acquired.[52]

RIBOT AND PSYCHOLOGICAL HEREDITY

Ribot's massive doctoral thesis (1873) on psychological heredity went through eleven French and ten American editions.[53] The cardinal principle

49 Fancher, "Alphonse de Candolle, Francis Galton, and the Early History of the Nature-Nurture Controversy," esp. 342.

50 Candolle, *Histoire des sciences et des savants* (1987, reprinted from 1885 edition); see also same title, 1873 edition.

51 Candolle, *Histoire des sciences* (1873) 159, 195–6, 289, 341, 414; Fancher, "Alphonse de Candolle," 345, 347.

52 Ribot, Review of Candolle, "Histoire des sciences."

53 Nicolas, *Théodule Ribot*, 17–23, 65.

was the importance of organic heredity – human predispositions in bodily constitution and nervous organization.[54] Somewhat like the present-day work of Steven Pinker, Ribot came out strongly against the empiricists' tabula rasa model. Heredity meant that "nature copies and imitates itself, this is a law of nature but also a law of change."[55] He based his ideas on both Lucas's medical treatise and Darwin's description of variation in domestic animals and plants. Therefore, heredity could also be a force responsible for variation and human evolution, most notably in the proportions of the nervous system and the dimensions of the brain (by inheritance of acquired characteristics). The civilized became more intelligent, even if they were not necessarily more moral. Organic unconscious memory inherited from ancestors disposed people to act as they did.[56]

In general Ribot supported Lamarckian inheritance of acquired traits, though he thought three or four generations might be needed to fix these traits. Contrary to some neo-Lamarckians who opposed including mutilation in these traits, Ribot believed that Inuit dogs would become tailless if their tails were removed for several generations. He also thought induced epilepsy in guinea pig parents would be passed to the next generation.[57] The result was a possibly optimistic interpretation for humans. In the same moral milieu, more moral education would fix beneficial characteristics. However, Ribot believed Lamarck gave the milieu too great a role, since the qualities of character preceded the variations, much as the phrenologist Franz Joseph Gall argued that the traits of his phrenological subjects pre-existed. Moreover, in wartime the influence of education combatting savage nature fell away like the scales of a fish.[58]

While the milieu intervened in the inheritance of acquired characteristics, Ribot insisted that "the biography of most famous men shows that the influence of education on them is sometimes nil, sometimes harmful, most often weak." So the "influence of education is never absolute and only has effective action on average natures," with minimal influence on the idiot or the genius.[59] In other respects, Ribot was always opposed to August Weismann's refusal to admit inheritance of acquired characteristics, and believed that revisions in Weismann's views meant that the arch-critic himself was open to the idea of inherited predispositions (in the germ plasm itself).[60]

54 Bénichou, "Ribot"; Ribot, *L'Hérédité psychologique* (hereafter *HP*), 269–70; 288.
55 Ibid., 1.
56 Ribot, *HP*, 6; Gasser, "La Notion de mémoire," 306–7; Otis, *Organic Memory*, 1–2, 10.
57 *HP*, 11.
58 *HP*, 327–9, 331.
59 *HP*, 328–9.
60 *HP*, i-ii.

Like Darwin and Spencer, he believed that moral and social instincts affected the survival of tribes. The qualities of sympathy, self-control, and duty, reinforced by education, customs, and religion, would be transmitted by heredity to descendants and would be forgotten only by atavistic characters.[61]

Heredity was otherwise responsible for transmitting organic sense, memory, and general intelligence.[62] Using Galton's lists of talented artists, Ribot insisted that heredity was the primary element in such talent, despite a role for education and tradition. He felt his methods were superior to Galton's, because they were more qualitative and did not rest on statistics.[63]

In the first edition of *L'Hérédité, étude psychologique*, but only in the first edition, Ribot even endorsed the idea that each individual should take responsibility for his own heredity and thus "there would be a natural means of eliminating from society the worst elements." Later, Ribot would show no interest in eugenics.[64] A factor qualifying optimism about heredity in later editions was the tendency of noble families or dynasties to degenerate, as physician Paul Jacoby had pointed out.[65] Ribot conceded to Candolle that physical dispositions were more likely inherited than instincts or moral dispositions, and moral dispositions, though a recent veneer, were more likely to be inherited than intellectual dispositions. But that did not lead to a change in the general argument. Without direct reference to Galton's twin studies, Ribot argued that twins often have the same tastes, faculties, and destiny.[66]

The inheritance of acquired nervous and mental disease was a trump card that Ribot often played against opponents of Lamarckianism. He also believed that criminal tendencies and passions such as greed or gambling ran in families. He cited Moreau de Tours and Morel in showing the transformations from mania and melancholy or epilepsy to idiocy and imbecility.[67] He noted that others such as Henry Maudsley estimated that only one-quarter to one-half of mental illness was inherited, though Ribot still valued the work of Moreau de Tours. He had to admit that heredity could be blocked by chance – accidental causes at conception or counteracting influences from parents. As

61 *HP*, 336–8, 341.
62 *HP*, 59, 66, 91–2.
63 *HP*, 71–2, 75–6, 115, 120, 217–18, 220.
64 Ribot, *L'hérédité, étude psychologique* (Hereafter *HEP*), 382.
65 *HP*, 276; Jacoby, *Etudes sur la sélection.*
66 *HP*, 175–6, 282, 329.
67 *HP*, 94, 246–7.

Lucas believed, there was a principle of diversity, an "innateness," producing difference, not resemblance.[68]

PSYCHOLOGY OF PEOPLES AND RACES

In the study of heredity, Ribot placed a great deal of credence in the importance of "national character" as the explanation of the moral and historical identity of a people. He wrote, "Institutions are the effect of customs and religious beliefs, that are in turn the effect of the character of a people ... that is, the habitual manner of thinking and feeling." While character and institutions influenced each other, character had an internal cause.[69]

Aside from Taine's efforts to explain the psychology of peoples, a "psychology of peoples" had already been promoted in the German states by the philosophers and philologists Moritz Lazarus (1824–1903) and his brother-in-law Heymann Steinthal (1823–1899) through their *Zeitschrift für Völkerpsychologie und Sprachwissenschaft* (1859).[70] Both authors envisaged "people" as a cultural community of language, myth, and customs. The Volk to them was subjective, more like a nation and unlike the anthropological concept of race. Ribot would attack the earlier German ethnic psychologist Theodor Waitz for asserting that climate, rather than innate differences, was the chief distinguishing mark among the human races. He also maintained that the periodical of Lazarus and Steinthal had produced no valid scientific generalizations from its study of language, religion, myth, and customs. Even in 1876 Ribot admitted, though, that "psychology plunges its roots in the life sciences and flowers in the historical sciences."[71] In France, Gustave Le Bon, Alfred Fouillée, and Emile Boutmy were developing the concept of constant psychological characteristics of a people. Le Bon believed that they were as stable as anatomy and inherited, while Fouillée thought that the weight of the social milieu would be more important in recent times.[72]

Ribot, however, uncritically used Paul Broca's figures on increased brain capacity in modern humans since the Middle Ages and those on

68 *HP*, 155, 234–5.

69 *HP*, 32, 120; see also Nicolas, *Théodule Ribot*, 57.

70 Diriwächter, "Völkerpsychologie"; Kalmar, "The Völkerpsychologie of Lazarus and Steinthal"; Trautmann-Waller, *Aux origines d'une science allemande de la culture.*

71 Ribot, *La Psychologie allemande contemporaine*, 40, 45–7, 52, 56; "La Psychologie ethnographique en Allemagne," 601.

72 Mucchielli, "Psychologie des peuples, races, regions et milieu social," esp. 84–5, 90–4 on Alexandru Dimitrie Xénopol versus Paul Lacombe in 1901–02; see also Paligot, "La Psychologie des peuples et ses applications durant l'entre-deux-guerres."

nineteenth-century differences between Australians and Europeans. These were intended to show both the effect of education on brain capacity and the significance of brain capacity for intelligence.[73] The difference among "racial" types showed the preponderant influence of heredity. The inheritance of acquired characteristics could explain increase in brain size.[74] Among the "European, Negro, Mongolic" human types the "intellectual, moral, and social differences (the only ones important to us here) are great, and their contributions to the work of civilization very unequal." The "Negro," for example, inherits sensitivity and imagination, sensual tendencies, and inability to abstract.[75] The "intelligence of the Negro appears incapable of passing a certain degree," and young Maori can never equal English children. Nor can Australians' limited intelligence fathom "complicated moral relations" such as justice, sin, crime, generosity, pity, and mercy.[76]

Ribot also adopted the anthropological commonplace that an "inferior race" represented the childhood of the species. Consequently, "savages" have an "incurable taste for adventure, [and an] inability to accommodate to our mores." Maori cannibals and primitive Filipinos would regret their old lifestyles if introduced to civilization. Like the anthropologist Charles Letourneau, Ribot thought that savages could not quickly achieve civilization. For once returned to a previous life, the savage does not retain or transmit civilized habits. Hence, there was difficulty in adaptation by a neo-Lamarckian mechanism, barring the transfer of whole tribes.[77]

On the characteristic anthropological quandary of the consequences of racial crossing, Ribot was initially suspicious of Arthur de Gobineau's fears about racial purity. In the first edition of *L'Heredité*, he praised the mixture of "disparate" Persian and Arab races that spread a great civilization with racial mixing, and thought that Jews blended elements from Syria, Persia, Phoenicia, and Greece. Moreover, the botanist Alcide d'Orbigny had commented that the Polynesians represented a flourishing mixed civilization.[78] But in later editions he began to fear the degradation or extinction of pure races. He preferred the analysis of polygenists J.-A.-N. Périer and Eugène Dally of the Société d'anthropologie and the Swiss-American Louis Agassiz who disparaged mixing to other anthropologists who were more tolerant, such as Martin de

73 *HP*, 302.
74 Ribot, *HEP*, 450; Nicolas, "L'Hérédité psychologique," 315.
75 *HP*, 167–8, 302, 344.
76 *HP*, 309, 312.
77 *HP*, 53, 95, 99, 299, 319.
78 *HEP*, 126, 413, 415.

Moussy, Armand de Quatrefages, and the German folk psychologist Waitz.[79] He thought that colonies were more prosperous and stable when conquerors avoided mixing with indigenous people, and found places like Haiti, Santo Domingo, and other Latin American countries unstable, with atavistic traits in mixed races.

The predominance of inherited traits also resulted in handicaps for apparently pure races. Basque gypsies, for example, had a "native incapacity transmitted hereditarily to adapt to civilized life" and would slowly perish.[80] Although climate and milieu did make a difference, he approved the labelling of Jews by Americans Josiah Nott and George Gliddon as possessing identical traits throughout the world. While praising their religious vocation, and their poetic and musical talents, he then proceeded to include a lengthy quotation from Semitic linguist Ernest Renan concluding that the Semitic race had no plastic arts, rational science, philosophy, political life, or military organization, and therefore manifested an incapacity for discipline and organization differentiating them from Europeans. In the later editions, he softened this condemnation (which so aggravated Adolphe Franck at the thesis defence) by using Candolle's lengthy demonstration about the pre-eminence of heredity, not religion, among Jews. A society of Jews, argued Candolle, because of hereditary habits, would be hard-working, intelligent, thrifty, sober, non-violent, wealthy and philanthropic, with stable marriages, but would be pillaged by their more violent neighbours.[81]

Thus he saw a limit to individuality and individual liberty. "Races themselves preserve their psychic as well as their physiological characteristics" and even varieties of races called peoples "present permanent moral characteristics, if one considers the totality of individuals."[82] Like Galton, however, Ribot thought that endless progress would not be possible even though education assisted the improvement from savagery to civilization. There would always be a regression to the mean. However, the difference between the brain of the savage and the brain of the civilized individual was infinite.

RIBOT'S RECEPTION

Serge Nicolas has discussed the reaction to Ribot's doctoral thesis.[83] Hippolyte Taine praised the work in the *Journal des Débats*. The provincial

79 *HP*, 344–6; Blanckaert, "Of Monstrous Métis?".
80 *HP*, 100, 136.
81 *HP*, 127–32, esp. 129–30.
82 *HEP*, 538–39; *HP*, 267, 412, 414; Nicolas, "L'Hérédité psychologique," 340.
83 Ibid., 318–20, 323–4, 327–8, 340–1, 343.

philosophy professor and left-wing Gambettist deputy Paul-Amand Challemel-Lacour praised its scientific sophistication, unhampered by clericalism, but thought the thesis defence had papered over disagreements with the committee. Fernand Papillon, philosopher and chemist, in the *Revue politique et littéraire* noted that Ribot was forced to retreat from any strictly associationist or materialistic pronouncements that would deny the activity of the self. He believed Ribot did not adequately acknowledge a world of spontaneity and liberty.[84] Both examiners, the Sorbonne philosophy professors Paul Janet and Elme Caro, had argued that the personality and will of the self win out against physiological tyranny.

Caro later presented Ribot's thesis to the Academy of Moral and Political Sciences.[85] Caro felt that Ribot's factual evidence for heredity was as well explained by milieu and education, and that Ribot did not endanger spiritualism. Rather, he confounded hereditary traits of the species with individual heredity. Adolphe Franck commented that heredity was nowhere a major influence, not even "pretended innate aptitudes in certain races," and both truth and morality would be violated if it were. In a more extended periodical article, Caro contended that negative counterexamples of the influence of the moral and intellectual atmosphere or of education on a child might be more telling than the alleged inheritance of scientific aptitudes. He objected less to the national character section but thought it was difficult "even for races, to determine rigorously these stable and permanent traits!"[86] There could also be efforts to change group character by imitation, not to mention the possibility of group crossing, which would "diminish the influence of national character." Still, in the races, the "individual element tends to be erased more and more to allow the reappearance of nature, that is, the species." So Caro did not rule out the importance of race, but questioned its stability and therefore was far more ambivalent than Ribot about assigning it importance.

The neo-Kantian and critical philosopher Charles Renouvier took strong exception to Ribot's seeming endorsement of hereditary determinism. He argued that "Mediterranean" parents could produce offspring inferior in intelligence to the average "Negro," and "ordinary parents can have a man as far above the ordinary as the ordinary man is superior to the ape."[87]

Surprisingly enough, one would expect that Francis Galton would have been an enthusiastic supporter of an ally in propounding the

84 Papillon, Review of Ribot, 1214–18.

85 Caro, "L'Hérédité psychologique, par Th. Ribot."

86 Caro, "L'Hérédité psychologique," esp. 55, 60.

87 Renouvier, "Des rapports du physique et du moral," esp. 314.

importance of heredity. However, as Galton's review of Ribot's work in English translation demonstrates, this was anything but the case. Claiming that heredity was now a science needing only experimental confirmation, he asserted that Ribot's book "was not on a level with present knowledge" and harked back to the pre-scientific stage of writing about heredity using questionable anecdotes. He thought Ribot too timid on proofs of heredity of memory. Most irritating, Ribot used Galton's tables and genealogies "appropriated without a word of acknowledgment." In fact, the English edition only acknowledged Galton's men of science and discussed Galton's statistics. There was no hint of the source of genealogies of other famous people. Galton concluded, "I laid down M. Ribot's volume after honestly reading every line of it, with a weary sense of many wasted hours."[88]

In 1884 the physiologist Charles Richet reported to the Société d'anthropologie on a commission, which included Ribot and Letourneau, on the desirability of a questionnaire on psychological heredity using Galton's "precious materials" as a precedent. As previously noted, Richet pointed out that "heredity and milieu are not antagonistic, but synergistic."[89] The society would probe physical resemblances, such as acuteness of senses and muscular strength, and psychic resemblances, such as memory and intellectual stamina, to parents and other relatives. They could then determine the preferential transmission of certain characteristics under specified conditions.[90] They hoped for some 1,200 to 1,500 replies to some 78 questions, but as Ribot later said, the replies were so vague and incoherent that they gathered dust after 1887. Years later, Ribot doubted whether questionnaires, as contrasted to tests, were a fruitful tool for psychological analysis, since for the public at large responses might be poorly formulated or derived from commonplaces. He thought they would work only for simple questions and large numbers.[91] Apparently, Charcot's Société de psychologie physiologique unsuccessfully attempted a similar project on psychological heredity. Candolle hoped that the questions would be addressed to mature people whose characters were well settled, and admitted the importance of understanding educational background, the social condition of parents, the subject's

88 Galton, Review of Ribot, *Psychological Heredity*; a check of *Psychological Heredity* (New York: D. Appleton, 1889), unchanged from the first edition, indicates that Ribot's acknowledgments to Galton, 72, and in the discussion of his statistics, 186–91, are few; cf. more extensive references in the later French editions, *HP*, 72, 77, 81.

89 Archives de la Société d'anthropologie de Paris, Musée de l'homme, 1K3, 6 mars 1884.

90 Richet, "Rapport sur un projet de questionnaire d'hérédité psychologique."

91 Ribot, "De la valeur des questionnaires en psychologie."

education, and the physical and moral state of the parents at the time of conception – all in an attempt to separate nature and nurture.[92]

RIBOT'S CONCESSIONS ON INHERITANCE

Yet for all of Ribot's fervor and his continual publication of new, virtually unaltered (after the second edition) versions of his doctoral thesis, his article on "heredity" for the Grande Encyclopédie in 1894 made major concessions on the inheritance of intelligence, if not of perception and memory. While he still supported Galton's dynasties of special talents such as mathematics or music, he now admitted that inheritance of intellect was "so complex that one cannot expect a rigorous transmission." Milieu may influence even the expression of elementary faculties. The son of a naval captain may not command a ship, but rather be a master of dogs in the hunt.[93]

But Ribot's turning point was by no means definitive. In the same article, Ribot was confident about the "inheritance of sentiments and passions." He thought that the alienists had satisfactorily demonstrated that nearly half of mental illnesses are inherited and he retained Morel's model of degeneration via heredity. He also made no concessions from an anthropological viewpoint on national character. Like Le Bon, he thought that one "could find from the infancy [of a nation] ... a collection of qualities and flaws of a purely psychological order ... that persisted with tenacity throughout the course of its history." Traditions, education, and customs constituted an enormous power for conserving habits, but "there are manners of feeling and thinking that are irreducible. True characters change no more among peoples than among individuals."[94]

There are several possible reasons for Ribot's more nuanced positions. First, Galton himself had admitted that intelligence is not as strongly inherited as other simpler qualities. Second, the German zoologist August Weismann's new theory of continuity of germ plasm had shaken, but not forced Ribot to surrender, the notion of inheritance of acquired characteristics. Hence, the milieu might operate more independently of inheritance of the characters it affected. Finally, the emerging discipline of sociology was proving to be a serious rival to the anthropological perspective. The young lawyer René Worms, who had discussed Alfred Espinas's work on *Animal Societies* with Ribot, founded a new *Revue internationale de sociologie* in 1893 with Ribot listed as a collaborator on the

92 Candolle, "Lettres sur un questionnaire d'hérédité psychologique."
93 Ribot, "L'Hérédité." In A. Berthelot, et al., *La Grande Encyclopédie*, esp. 1166.
94 Ibid.

editorial board. Though Worms stressed analogies between society and biological organisms, the institutions he formed debated and challenged simplified biological determinism.[95] Ribot had read Emile Durkheim's *The Division of Social Labour* (1893), which argued that, in more developed modern societies, heredity transmitted only the most general aptitudes and predispositions, rather than specialized individual traits. Moreover, family circumstances, education, and wealth were more likely to affect these aptitudes than hereditary influences.[96] Ribot's *Revue philosophique* also first published Durkheim's *Rules of Sociological Method*, which insisted on explaining social facts by other social facts.[97] Durkheim would continue this critique of heredity in *Suicide*, while his colleague Célestin Bouglé began an onslaught on organicism and Vacher de Lapouge's anthroposociology.[98]

Nevertheless, Ribot's willingness to defend the influence of heredity and to expand its role through inheritance of acquired characteristics did not weaken. He believed Weismann had no refutation of the transmission of acquired nervous and mental diseases. Nor could Weismann refuse to admit that certain predispositions were inherited, even if external influences provoked variations. Ribot would still rather believe that parents' efforts were worth something by way of inheritance to future generations. Weismann offered no cause of the variability he admitted, since he would not resort to external environmental sources of variation of germ cells.

CULTURE OR RACE?

Two years after the encyclopedia article appeared, Ribot seemed to signal a substantial turning point, assigning full credit to cultural influence. In the preface to the *Psychologie des sentiments* (1896) he now admitted that affective life cannot be separated from "social, moral, and religious institutions ... if the life of the mind has its roots in biology, it develops only in social facts." So psychologists needed sociological sophistication in explaining the development from primitive cannibalism to the present moral and social culture, from a beastly sexual act to chivalric love,

95 Discussion of "Théorie organique de la société," *Annales de l'Institut international de sociologie* 4 (1897): 169–341; *Revue internationale de sociologie* 1 (1893) title page; Lenoir, "Lettres de Théodule Ribot à Alfred Espinas (1876–1893)," esp. 172; Mucchielli, *La Découverte du social*, 270–5.

96 Durkheim, *De la Division du travail social*, 291–318; Lenoir, "Lettres," 172.

97 Durkheim, "Les Règles de la méthode sociologique," *Revue philosophique* 37 (1894): 465–98, 577–607; 38 (1894): 14–39, 168–82.

98 Hecht, "The Solvency of Metaphysics."

or from gross fetishism in religious metaphysics to spiritually sophisticated mysticism. The "essential, fundamental cause is intellectual development" while another cause, "more doubtful and limited in its action, is hereditary transmission."[99]

Yet Ribot had already asserted a similar need for the social sciences beyond psychology in 1877, and even now he barely emerged from the anthropological shadow. The same book still asserted the "known fact of inequality of human races in abstracting and generalizing" when looking at the evolution of languages, the scientific spirit, or the history of religion. Some peoples remained capable only of forming general simplified and condensed general images.[100] In discerning a trend in the history of religion, Ribot used the research of British cultural anthropologist Edward Burnett Tylor and French historian of religion Albert Réville. In savage polytheism or polydemonism, each tree or each forest had a god, while higher-level religions had water, fire, and earth gods. At a more advanced stage of abstraction, there were national gods, and finally, a universal God, who represented a conquest of the metaphysical spirit, not popular intuition.[101]

Ribot's study of the evolution of general ideas consigned inferior sorts of abstraction, as contrasted to the moderate and superior kinds, to animals and children, deaf-mutes, and the "languages of primitives and half-civilized peoples."[102] The Bushmen, for example, needed action language along with their speech, so that they could not communicate at night except around firelight. And, as if to deny the conclusion that only inferior culture was involved, he continued, the "inability of inferior races to do the most simple calculations is well known."[103]

In the affective sphere, those closest to nature and the lowest on the evolutionary scale were the angriest. Children's behaviour was aggressive toward animals, the weak, and the defeated, as was the conduct of primitives, gross people, and the mentally challenged. Savages laughed brutally in victory.[104] Removing for the sake of argument the influence of heredity, Ribot noted that sentiments obviously changed under the influence of education, tradition, and the social milieu, as even Weismann and Alfred Russel Wallace would have to acknowledge. Yet Ribot could not bring himself to disavow the Spencerian inheritance

99 Ribot, *La Psychologie des sentiments*, vii-viii, 186–7; Mucchielli, "Aux Origines de la psychologie universitaire," 288.

100 Ribot, *Psychologie des sentiments*, 18, 305.

101 Ibid., 305–6; Ribot, *Evolution des idées générales*, 107–10.

102 Ibid., 2–3.

103 Ibid., 78, 111, 113, 234.

104 Ribot, *Psychologie des sentiments*, 219, 345.

of acquired appetites, tendencies, and passions, whether fear, benevolence, or love of nature.[105]

In some passages, Ribot tended to discuss culture more than race. He was willing to concede that altruism and social sentiments existed among "primitives," since Tahitians and Fijians abolished cannibalism on their own. These social sentiments came from maternal love, attraction, and gregariousness. But a highly refined morality needed a more advanced cultural context: " A Saint-Vincent de Paul is as impossible among the Canaques [New Caledonians] as a Mozart among the Fuegians." In a certain cultural milieu, instinctive sympathies would not be sufficient for refined morality.[106]

In his critique of criminal anthropology, Ribot cited the alienists Prosper Despine and Henry Maudsley to conclude that moral blindness in individuals was "most often innate and coexists with other symptoms of degeneracy." Moral blindness was instinctive and affective, not an intellectual activity – rather, an inherent defect of character.[107]

So Ribot made concessions on the heritability of intellect and qualified the all-powerful nature of heredity in affective life in the face of the critique of inheritance of acquired characteristics. But a concern for social milieu and culture did not entirely remove the focus on the quality of the race. Moreover, the anthropological commonplace remained – primitive cultures would need a long period of time to attain civilized refinement.

In 1900 Ribot even more clearly identified the highly developed imagination shown by myth-making in "primitive man" as a "fixed state" of arrested development, never to be surpassed, as it is among civilized people.[108] Concurrent influences affected myths: the "characteristics of the race," an inherent imaginative power, the "style of life," savage or civilized, and the "cosmic milieu" of the natural environment. In "well-endowed races," there would be a growing power of observation and reflection.[109] So both race and milieu affected myth-making and contributed to potential for advancement. For each talent, from music to science, education was needed for its development. At this stage Ribot had very much anticipated later views that the environment is responsible for the expression of certain genetic endowments. But if milieu alone acted, the unique endowment of the individual would be irrelevant.[110] And in

105 Ibid., 235.
106 Ibid., 290–1.
107 Ibid., 296, 418.
108 Ribot, *Essai sur l'imagination créatrice*, 93–4, 99–100.
109 Ibid., 107, 112.
110 Ibid., 122, 125.

his most generous statement toward "primitives," in contrast to his colleague Henri Joly, who said "superior minds" exist among them only in relative fashion, Ribot commented that the only way we can interpret "superior minds" in any context is in relative fashion, since "to beings better endowed than us, our great minds would look small."[111]

In his long career, Ribot eventually became more willing to stress the importance of social rather than biological factors in human development. However, in his 1909 survey of the "comparative" psychological method, he advocated a hypothesis of linguistic stages of development paralleling intelligence. Languages (monosyllabic, agglutinative, and flexion-based, in Max Müller's famous classification) were the "petrified psychology of a people." They exemplified precision or vagueness, rationality or emotion, and showed that "not all members of humanity were made in the same mould." Among primitive peoples, there was an "unequal development of their faculties where impulse dominates."[112]

In the *Logique des sentiments* (1905) that was so pleasing to Emile Durkheim, Ribot certainly manifested an awareness that the prejudices of myths and beliefs that he called a separate "logic of sentiments" were how human groups were "formed and maintained."[113] The key to superstition and magic, according to Edward Burnett Tylor, was the understanding of the subjective component, the mistaken deductions in emotional reasoning.[114] Although Ribot did not discuss these questions from a sociological viewpoint, Durkheim saw some resemblances to his own work on the logic of classification. He wrote an enthusiastic review in the *Année sociologique* praising Ribot for going as far as a psychologist can in discussing these phenomena. Ribot himself wrote to Henri Piéron that he conceived of the sociological side of the question too late and was willing to leave it to sociologists.[115] Yet unlike Durkheim, Ribot maintained a strong belief in a different mental constitution of "primitive man" or "present-day savages" – the "very inferior level of their logical faculties, inaptness for abstraction, extreme difficulty of linking ideas according to objective relationships."[116] While the logic of sentiments

111 Ibid., 127.

112 Ribot, "Les Méthodes de la psychologie," in F. Thomas, ed., *De la Méthode dans les sciences* (Paris: Alcan, 1909) republished (2000) in Marchal and Nicolas, "L'Emancipation de la psychologie," *Psychologie et histoire* 1 (2000): 43–81. http://psychologieethistoire.googlepages.com/MARCHAL1.HTM, original 243, 245.

113 Ribot, *La Logique des sentiments*, 17.

114 Ibid., 19.

115 Durkheim, Review of Ribot, "La Logique des sentiments," 156–8; Letter of Ribot to Piéron in Carroy and Plas, "La Psychologie: Science naturelle et science morale?" 342–3; Mucchielli, "Psychologie et sociologie en France, l'appel à un territoire commun."

116 Ribot, *La Logique*, 38, 176.

would never disappear in the modern world, the logic of rationality could not yet appear among primitives. With the logic of sentiments, the principle of contradiction did not necessarily take hold, hence myth-making could precede the differentiation of types of logic.[117]

CONCLUSION

Fascinated by French and British studies of hereditary disease, enamoured with a Lamarckian belief in the tendency to inherit acquired organic dispositions, and furnished with empirical ammunition from Galton's family dynasties, Ribot magnified the role of heredity in a truly scientific psychology as well in the physical composition and functioning of the body. He had been enthralled with the hard-line physicalist anthropological perspective to the point of criticizing German efforts at formulating a psychology of people based on culture. Yet the debate over Weismann's theory of continuity of germ plasm forced him to question the neo-Lamarckian inheritance mechanism – a questioning that could have opened up more room for the independent action of the milieu apart from inherited transmission. However, he never surrendered the Lamarckian position. The intellectual world around Ribot was not static during his long career. Emile Durkheim published his *Rules of Sociological Method* under Ribot's editorial supervision of the *Revue philosophique*, and sociological viewpoints concerning the role of religion or institutions increasingly infiltrated Ribot's own psychological treatises.

Scholars such as Mucchielli have cited Ribot's posthumously published contribution to Georges Dumas's treatise of psychology as the culmination of a trend in Ribot's thinking toward openness to the sociological. Ribot would state, "if psychology begins with biology and zoology, it flowers finally in sociology," so that "psychologists must recognize the original character of the social fact."[118]

Yet it appears that Ribot's concessions to the importance of the social fact were half-hearted. His refusal to surrender belief in the inheritance of acquired characteristics gave an important biological, hereditary component to the activities shaped by the milieu. While Ribot emerged from the shadow of physical anthropology, he retained the assumptions of cultural anthropology. The inequalities of intelligence and character were due to a "primitive" stage of cultural development, if not to racial aptitudes. Since some peoples obviously achieved civilization while others did not, his views were consistent with the anthropological assumptions of limited perfectibility for some peoples. Even the most sociologically oriented

117 Ibid., 63, 97.
118 Mucchielli, "Aux Origines," 289.

thinkers such as Emile Durkheim himself were not free of all assumptions of racial hierarchy. But Ribot never really wrote consistently as if he believed sociology was as important as the organic and hereditarian perspective. Beyond Ribot's own trajectory, however, the articles he helped publish in the *Revue philosophique* expressed the general trend away from physical anthropology and toward a more sociological viewpoint.

5

Heredity and Milieu in the Revue philosophique

WHILE THE *REVUE PHILOSOPHIQUE* encompassed every conceivable area of philosophical interest, Ribot had a special concern for the issue of psychological heredity. Laurent Mucchielli has studied the origins of the periodical, in which Ribot sought to defend doctrines of evolutionism, experimentalism, and naturalism with a network of fellow Ecole normale graduates, evolutionist psychologists, and physicians.[1] My own analysis here of over 180 articles and reviews concerning heredity and milieu in the *Revue philosophique* reveals an increasing displacement of an anthropological by a sociological perspective that privileged milieu over heredity. However, even advocates of the milieu were not immune from a European superiority complex.

Opinions on gender were often less flexible than those on race, and there was no obvious correlation between views on race and views on gender. In the very first issue of the *Revue* Ribot published articles by the most influential promoters of scientific psychology, the anti-spiritualist Hippolyte Taine and the evolutionist Herbert Spencer. Most reviewers stressed their ideas of reasonably stable racial attributes, even though they both clearly allowed a major role for milieu. Only a few philosophers, such as Jean-Marie Guyau, argued for the enormous power of education. Authors such as Ribot's friend Gustave Le Bon and the Ecole d'anthropologie lecturer Charles Letourneau, however, represented an anthropological approach that made either heredity all-powerful or cultural retardation almost insurmountable. Even in the 1880s the previously discussed anthropological countercurrent foreshadowed the "crisis

1 Mucchielli, "Aux Origines de la psychologie," 263–89, esp. 271, 275–82; for other views of the periodical see Thirard, "La Fondation de la 'Revue philosophique'," 401–13; Bertolini, *Il pensiero e la memoria.*

of anthropometry": the doubts about whether physical measurements justified conclusions about intelligence and character.[2]

Nuanced views on heredity and milieu also emerged in the works of leading philosophers, such as Alfred Fouillée, and pioneering sociologists, such as the magistrate Gabriel Tarde and founder of the Paris Société de sociologie, René Worms. Vocal criticism of hereditarianism appeared increasingly after 1900 in the new sociological movement of Emile Durkheim, Gaston Richard, and Célestin Bouglé. But some philosophers and physicians were still reluctant to abandon racial hierarchy. The importance of uncompromising hereditarianism was being challenged, but the debate was far from closed.

TAINE, SPENCER, AND GUYAU

A precedent for Ribot's opinions on "primitives" appeared in Hippolyte Taine's contribution to the first issue of the *Revue*. Taine's article on language acquisition in children compared the mental state of the child to the poetry and mythology of the mental state of "primitive peoples."[3] However, the physiological underpinning of brain size and race was clearly evident. For Taine, races with less cerebral development could not demonstrate the "greater aptitude for general ideas" of Greeks and Aryans. In appraising Taine, Ribot fervently hoped for a "science of races" "cultivated by well-endowed minds used to scientific procedures" for the "innate element of race … fixed by heredity" – an ethnographic psychology superior to the Germans' inadequate "psychology of peoples."[4]

By the early twentieth century, Taine evoked criticism in the *Revue* as well as residual loyalty. A member of the Paris Société de sociologie, Arthur Bauer, charged that anthropologists failed to account for the modifications of sensibility, character, and aptitudes due to "milieu, education, especially style of life." The physical-mental relations they affirmed were often coincidences. Bauer said that even Taine, who conceived of race as an absolute and immutable entity, should have realized that nations had no stable moral character. For Bauer, the study of social classes gave a better indication of the state of society.[5]

A second critic of Taine, the historian and child psychologist Paul Lacombe, asserted that, despite the existence of physiologically diverse races, the "power and effectiveness attributed to race does not exist."

2 Blanckaert, "La Crise de l'anthropométrie," 95–172.

3 Taine, "Notes sur l'acquisition du langage chez les enfants et dans l'espèce humaine," 5–23, esp. 14.

4 Ribot, "M. Taine et sa psychologie," 19–21, 46.

5 Bauer, "Classes sociales," 292–3, 296.

First, race "is perpetually in a state of becoming" due to migrations, intermarriage, and new political associations. Second, the social milieu assumed greater importance than the physical milieu or race. The reviewer, the neo-critical philosopher Frédéric Paulhan, still defended Taine's preface to the *History of English Literature* as indicating profitable directions for exploring the respective contributions of race and milieu.[6]

Herbert Spencer's comparative psychology also showed strong allegiance to the anthropological commonplace that larger brain volume and healthful physiology produced intellectually superior races. Like Taine, he maintained that childlike inferior races could not generalize or focus attention. However, Spencer also argued that diet and lifestyle influenced the character of a society. He was receptive to correlations of social conditions and institutions with mental development. Spencer believed in a cycle of biological and cultural inheritance, with a blurring of distinctions.[7]

Spencer believed in the hereditary transmission of "moral intuitions" from past generations.[8] Surprisingly Ribot's friend, the educator Gabriel Compayré, criticized Spencer's insufficient attention to psychological heredity.[9] A more common reading of Spencer found his racial hierarchy too clearly fixed. Jean-Marie Guyau, the young moral philosopher who died prematurely at the age of 34 in 1888, found that Spencer's alleged "moral intuitions" would justify aristocratic or racial prejudice. Chinese women, he noted, did not inherit the desire for narrow feet, nor Muslim women the desire to be veiled. In reality the "considerable increase of the influence of the physical and moral milieu" guided people. However, even Guyau believed the milieu could not guarantee that a Canak (New Caledonian) girl could be "raised with the same delicacy of sentiments or elevation of ideas" as a European girl.[10]

In 1879 Guyau had articulated a predominantly environmentalist position. Ribot's stalwart reviewer, the art historian Lucien Arréat, however, found Guyau too disdainful of psychological heredity. Inherited illnesses and degeneracy showed troubles in the nervous centres and

6 Paulhan, Review of Paul Lacombe, "La Psychologie des individus," 419–25; Lacombe, "Milieu et race," 39–40.

7 Spencer, "Esquisse d'une psychologie comparée de l'homme," 47–8, 57–8, 61; Carneiro, "Herbert Spencer as an Anthropologist," 174–7; Gissis, "Herbert Spencer's two editions of the Principles of Psychology," 137–52.

8 Guyau, "L'Hérédité morale de la théorie de Herbert Spencer," 308–9.

9 Compayré, "Les Principes de l'éducation d'après Herbert Spencer," 168–85, esp. 184.

10 Guyau, "L'Hérédité morale," 310–13.

transmission of a more profound mental disturbance to the next generation. Arréat believed individuals were born with moral capacities, and that education created habits fixed later by heredity in the race. In itself education could not be the primary factor in the evolution of human societies.[11]

ANTHROPOLOGY OF LE BON, LETOURNEAU, AND THE IMPORTANCE OF RACE

Besides Taine and Spencer, some anthropologists promoted hereditarian or weak environmentalist perspectives. In 1877 the eccentric anthropologist and psychologist Gustave Le Bon insisted that "successive hereditary accumulations" changed character, compared to the minimal influence of milieu and education.[12] The next year Le Bon summarized his cranial research, where he found that the larger number of superior-sized crania in "advanced" races was even more significant than differences in mean size among the races. Furthermore the small mean size for women led to conclusions about their reduced intellectual potential. The historian Charles Seignobos bristled at Le Bon's denigration of the abilities of women as if they were a different species unable to inherit the capacities of male ancestors.[13] Le Bon later argued that "slow change produced by variability of milieux, selection, and accumulated by heredity" differentiated the races. Here he articulated his argument of the weak present-day milieu and the invariability of a Jewish race regardless of milieu.[14]

The philosopher Frédéric Paulhan reviewed Le Bon's theory of stable "historical races." For Le Bon, inferior races like the Ethiopians degraded Egyptian art, but superior races like the Greeks highly refined it.[15] Paulhan did not question Le Bon's hierarchy of races in itself but doubted his pessimistic argument that intelligent peoples eventually lost the virtues of good character.[16] He also criticized Le Bon's more famous crowd psychology. There the author compared impulsive, suggestible crowds to the "characters of beings of inferior forms of evolution, such

11 Arréat, "Récents travaux sur l'hérédité," 407, 416, 419.

12 Le Bon, "Notes sur l'étude du caractère," 497, 499–500.

13 Le Bon, "Recherches expérimentales sur les variations de volume du crâne," 207–9; Seignobos, Review of Le Bon, 202.

14 Le Bon, "Applications de la psychologie à la classification des races," 594; Seignobos, "Les Premières civilisations," 199.

15 Paulhan, Review of Le Bon, "Les Lois psychologiques," 513–14, 516, 519.

16 Ibid., 519–22.

as the woman, savage, and child."[17] The reviewer reprimanded Le Bon for rejecting an upward trajectory of human progress aided by cultural diffusion. Clearly other *Revue philosophique* contributors were critical of Le Bon, though they often apparently accepted racial hierarchy even after the anti-hereditarian onslaught of sociologists. Le Bon's influence was evident in his many contributed articles and the many comments of reviewers on his books. He also demonstrated how strong hereditarianism had hierarchical implications for race and gender theory.

Charles Letourneau's evolutionist views that permitted racial progress still maintained that primitive races would take centuries to catch up with those more advanced. Once again Paulhan was critical of an overly biological slant in his sociology but did not attack the gradation of races.[18] By 1897 the rise of more confident sociologists cast doubt on Letourneau's credulity for travellers' tales. The Durkheimian Gaston Richard believed that Letourneau's description of various races from "lowest" to "most elevated" might be applicable to early historical epochs, but not to more recent eras, when civilization decreased racial influence. The transformation of slavery was more likely due to diverse civilizations than races, especially since Asians had no slavery while the allegedly more advanced "white" Muslim Semites still practiced it. Moreover, religions, more than racial origins, differentiated peoples such as the Mongols, the Turks, and the Magyars. Richard wondered, "Can we believe social development is thus subordinated to race?"[19]

Even after 1900 the hard-line physicalist view of race did not disappear. The moral and educational philosopher at the Poitiers Faculty, Marcel Mauxion, repeated the anthropological convention that primitive societies were like children without a sense of justice. He quoted Spencer to argue there was "more distance from a Papuan to Newton than from a chimpanzee to a Papuan." He even doubted whether "slavery, wisely understood and practiced with humanity" might not be necessary to impose "salutary discipline and rules of work" on "peoples with impulsive souls and lazy muscles." Equality was no inherent human right, and liberty was defined as the right of conquest.[20]

THE ANTHROPOLOGICAL COUNTERCURRENT

In 1877 the educational reformer and philosopher Henri Marion showed that Quatrefages, though not a transformist, persisted in his

17 Paulhan, Review of Le Bon, "La Psychologie des foules," 80.
18 Paulhan, Review of Letourneau, "La Sociologie d'après l'ethnographie," 546–51.
19 Gaston Richard, Review of Letourneau, "L'Evolution de l'esclavage," 644–7.
20 Mauxion, "Les Eléments et l'évolution de la moralité," 1–29 and 162–3.

belief in the importance of climatic differentiation of human varieties. As long as there are poles and an equator, continents and islands, or mountains and plains, "races will exist distinguished by characters of every kind, and superior or inferior in a physical, intellectual, and moral point of view." Two years later Quatrefages admitted, "the development of the intellectual faculties of man is, to a great extent, independent of the capacity of the cranium and the volume of the brain."[21]

Léonce Manouvrier came to symbolize the crisis in anthropometry. Manouvrier was not by disposition a pure egalitarian, as we have seen in his refusal to endorse Anténor Firmin's plea for equality for Africans.[22] I have also referred to the limits of his support for feminism. But militant anti-hereditarian sociologists such as Célestin Bouglé, discussed in the section on the rise of sociology, would find Manouvrier a kindred spirit in undermining the conventions of traditional anthropology and the "anthroposociologist" Georges Vacher de Lapouge's confidence in the cephalic index as a sign of intelligence.[23]

WEISMANN AND THE NEO-LAMARCKIANS

While anthropologists continued this debate, naturalists and biologists reappraised the very mechanism of heredity after August Weismann's challenge to the notion of inheritance of acquired somatic characteristics. Critics of Weismann argued that his "continuity of germ plasm" could not explain the adaptability of organisms to the milieu or the degeneration of organs by disuse. A naturalist and paleontologist at the Muséum national d'histoire naturelle, Félix-Edouard Bernard, found Weismann bold, but less rigorous than Darwin and Wallace.[24] Another common charge was that Weismann himself realized the implausibility of his theory and modified it. The zoologist from the Paris Faculty of Sciences Yves Delage noted that Weismann's more recent papers allowed that "determinants" of the germ plasm could alter it. If modifications over generations could accumulate to produce new, fixed hereditary characters, Weismann had in effect returned to Lamarck's ideas. Delage,

21 Marion, Review of Quatrefages, "L'Espèce humaine," 639–42, citation on 639; Quatrefages, *The Human Species* (Appleton), 284, 384.

22 Manouvrier, Review of Firmin, "L'Egalité des races humaines," 180–2; Hecht, "A Vigilant Anthropology," 221–40.

23 Manouvrier, "Les Aptitudes et les actes," 918–51; "L'Indice céphalique et la pseudo-sociologie," 233–59, 280–300.

24 Felix Bernard, Review of Weismann, "Sélection et hérédité," 646–56, esp. 656.

however, conflated inheritance of acquired somatic variations (which Weismann excluded) with acquired germ plasm variations.[25]

While the Lamarckian perspective was conducive to optimism about education and social change, the belief in transmission of acquired diseases was a pessimistic special case. The medical specialist Joseph-Jules Dejerine (1849–1917) was sympathetic to Weismann. But he strongly argued for the neo-Lamarckian possibility of acquiring through nervous exhaustion an inheritable "neuropathic" constitution, manifested as alcoholism, epilepsy, or hysteria.[26] Mental degeneration, with progressive deterioration, might occur when such conditions were present in both parents and in several generations. Hence neo-Lamarckians could explain the social ills of "degeneracy" as well as improvement of the species.

By 1913 the alienist and frequent reviewer Philippe Chaslin criticized the theory of degeneracy because it postulated "heredity of transformation" of one trait into another in subsequent generations. He also thought pathological phenomena could explain alleged degeneracy. For example, a spirochete caused spinal cord syphilis, once believed to be mysteriously inherited from parents. For Chaslin pathology was intrinsically distinct from heredity, since heredity was the reproduction of similar phenomena in the next generation, while pathology implied an alteration of the organism.[27]

On the whole the supporters of inheritance of acquired characteristics felt they had a trump card in pointing out congenital illnesses of this kind. There were far more critics than supporters of Weismann among the physician authors and reviewers. One of the most vehement neo-Lamarckians and prolific contributors was a biology professor and embryologist at the Paris Faculty of Sciences, Félix Le Dantec (1869–1917). He believed Weismann's explanations of variation inadequate.[28] Denying inheritance of acquired characteristics, he thought, would be like an African king denying that a trolley could run by electricity because he did not understand it. He also suspected that Hugo de Vries's mutation theory explained nothing.[29] With Weismann's assumptions animals would never adapt to meet their

25 Delage, "La Nouvelle théorie de l'hérédité de M. Weismann," 584; Winther, "August Weismann on Germ-Plasm Variation," 517–55.

26 Gley, Review of Dejerine, "L'Hérédité dans les maladies du système nerveux," 312–13.

27 Chaslin, Review of Génil-Perrin, "Histoire des origines et de l'évolution de l'idée de dégénérescence," 544–5.

28 Le Dantec, "Les Néo-Darwiniens et l'hérédité des caractères acquis," 1–41.

29 Le Dantec, Review of de Vries, "Espèces et variétés," 531–2; Review of Cuénot, "La Genèse des espèces animales," 642.

needs and would be at the mercy of chance.[30] Le Dantec, unlike others, also found Lamarck a bulwark against "genealogical pretensions" and the "voice of blood." Fortuitous biochemical combinations were the actual mechanism of hereditary transmission.[31] For Le Dantec, neo-Lamarckianism was a defence against aristocratic and racial ideologies. Conversely, conservative thinkers suspicious of science such as Ferdinand Brunetière and Paul Bourget saw Weismann as refuting the optimistic, egalitarian views of staunch republicans.[32]

Against the Mendelian Lucien Cuénot, a physiologist from the Nancy Faculty of Sciences, Le Dantec argued that representative characters corresponding to chromosomes would mean an arbitrary decomposition of the organism. Considering theoretical assumptions like Weismann's external influences affecting germ cells and Cuénot's argument for preadaptation of the organism, one might as well believe the creation of Adam and Biblical cosmogony. For the "accumulation of tendencies" was like an invisible spirit or jinni.[33] So neo-Lamarckianism was a scientific defence against crypto-spiritualism as well as against aristocratic pretensions. Clearly, both advocates and opponents of racial hierarchy could adapt a chameleon-like neo-Lamarckianism. For Ribot, neo-Lamarckianism had magnified the importance of race through ultimately fixed inherited characters, while for Le Dantec it was Weismann who represented the spectre of racial purity through the germ plasm.

Ribot's disciple, the psychologist Ludovic Dugas, also expanded the definition of heredity to include any phenomenon of contrast as well as of resemblance. He thought psychological heredity could be "heredity by contrast," the transmission of opposite characteristics.[34] With such an elastic definition of psychological heredity almost any phenomenon could be ascribed to it.

FOUILLÉE, TARDE, AND WORMS: INTERMEDIATE VIEWS

In addition to the anthropologists and naturalists, another general category of contributors to the *Revue* was philosophers or sociologists of a certain renown who took a solidly intermediate position on the issues of heredity and milieu. One of the most frequently published authors was the former Ecole normale philosophy lecturer (after Ribot's time) Alfred Fouillée (1838–1912). In 1891 he announced that cerebral and mental

30 Le Dantec, "Les Néo-Darwiniens," 14, 29, 36.
31 Le Dantec, "La Théorie biochimique de l'hérédité," 494.
32 Vatin, *Trois essais sur la genèse de la pensée sociologique*, 189–90.
33 Le Dantec, Review of Cuénot, 637, 640–2.
34 Dugas, "Une Hérédité psychologique par contraste," 276, 282, 284.

structure developed from both acquired habits transmitted by heredity and the direct action of the milieu on the brain or senses.[35] The acquired habits helped explain the diversity of individuals' moral and physical constitutions.[36] Later he argued that the social milieu had become more important in modern times, as physical factors gave way to social factors, and race became secondary.[37] While Fouillée rejected Le Bon's stability of character in "historical races," he also could not accept Durkheim's argument that no social phenomenon was dependent on race. Africans, he said, could not have conceived the art of the Athenians or the writings of the Hebrews.[38] Like Manouvrier, Fouillée thought the cephalic index was no reliable guide to civilization, and one had to reject arbitrary theories on psychology of peoples coming from anthropology, linguistics, or history. Still, cranial form had some importance, and, as Letourneau thought, savage peoples would take extended timespans to develop their brains.[39] The principle that race conditioned social and intellectual development, but was not the dominant factor in history, was also the theme of Fouillée's even broader survey of European peoples, where "historical races" or peoples were the primary object of study. There he insisted that an educable individual from an "inferior or degenerate race," that is a non-European, would always be limited by "native conformation" to a "development not surpassing certain limits."[40]

Ribot did much to encourage publications by the criminologist and sociologist Gabriel Tarde (1843–1904). Tarde also participated in René Worms's Institut international de sociologie in 1894 and was the first president of the Paris Société de sociologie in 1895. Tarde took a strong stand against overemphasis on race while opposing the minimizing of biological and especially of psychological factors.[41] For him "race is the daughter and not the mother of social facts, and by this forgotten aspect alone it appears to me to enter into the domain properly speaking of the sociologist." He also appeared to endorse a variant of Le Bon's historical races since he argued that from ancient times

35 Fouillée, "Les Origines de notre structure intellectuelle et cérébrale," 573.

36 Paul Souriau, Review of Fouillée, "Tempérament et caractère," 624–33.

37 Fouillée, "Les Facteurs des caractères nationaux," 5, 10.

38 Ibid., 11–12, 21.

39 Dugas, Review of Fouillée, "Psychologie du peuple français," 323–6; Paligot, *La République raciale*, 166–7.

40 Malapert, Review of Fouillée, "Esquisse psychologique des peuples européens," 438; Paligot, *La République raciale*, 155, 161; Grandmaison, *La République impériale*, 100–1.

41 For his critique of Durkheim, see Tarde, "Criminalité et santé sociale," 148–62; reply by Durkheim, 518–23.

habitual practices became customs, or "race instincts." However, he did not believe in the stability of these customs.[42]

Tarde, like Manouvrier, was a well-known opponent of the Italian idea of the "born criminal." Enrico Ferri, one of Cesare Lombroso's disciples, had attributed far too much to heredity, Tarde thought, and not enough to education in the broadest sense.[43] The criminal was not identifiable with the savage, nor was a prognathic face the exclusive attribute of particular races.[44] Criminality was not a natural condition, attached to facial features or cephalic index, but a relation to public opinion and legislation, more like belonging to an industrial corporation than like being a savage or a madman. No one was born to kill, steal, or commit arson.[45] In assessing the Lombroso school, Tarde maintained that most recidivists were not epileptic; hence a favoured explanation of the habitual criminal was invalid.

In 1890 Tarde pointed out that nations of the same race had completely opposed characters and that ethnic character changed over time. "There's no race that's not civilizable or barbarizable," and "no natural character that can't be made worse or improved, and in quite different ways, by the effect of education." Contrary to Le Bon, he thought Jews differed in different countries.[46] The varying attitude in the United States to African-Americans and Native Americans was evidence of aversion to slaves and their descendants, not merely a reaction to physical differences.[47]

Equally positioned squarely in the middle of the heredity-milieu debate was René Worms (1869–1926), the founder of sociological institutions such as the *Revue internationale de sociologie* (1893), the Institut international de sociologie (1893), and the Paris Société de sociologie (1895). Worms subscribed to the organic model for society but not to primacy of racial theory.[48] He insisted on the psychological nature of social facts and the biological origins of societies. Heredity and education,

42 Tarde, "Les Traits communs de la nature et de l'histoire," 270–91, esp. 282n.

43 Tarde, "Socialisme et criminalité," 517.

44 Tarde, "Le Type criminel," 595.

45 Ibid., 605, 613, 625; Borlandi, "Gabriel Tarde et les criminologues italiens de son temps," 7–56.

46 Tarde, "La Misère et la criminalité," 506–7.

47 Tarde, "La Logique sociale des sentiments," 579–80.

48 For Worms's institutions, see Terry Clark, "Marginality, Eclecticism, and Innovation," 12–27; Geiger, "René Worms, l'organicisme et l'organisation de la sociologie," 345–60; Mucchielli, *La Découverte du social*, 271–5; Barberis, "In search of an object," 51–72, esp. 56; Gissis, "Late Nineteenth Century Lamarckism and French Sociology," 69–122.

race and milieu, both acted on the individual.[49] Durkheim, he thought, was excessively mechanistic in ignoring the wills and purposes of individuals. Traditions acted like heredity in societies, which evolved like living beings, with social progress analogous to organic adaptation.[50]

While Montesquieu, Taine, the family sociologist Frédéric Le Play, and philosopher of social types Edmond Demolins stressed the action of milieu, Worms wished to consider the reaction of humans on milieu and their migrations.[51] He also hoped to include variables such as population density, facility of communication, social groupings, and, like Taine, the historical moment, as well as the interaction of great men and society.[52] Chapter 7 provides further discussion of Worms's views outside the pages of the *Revue philosophique.*

THE RISE OF SOCIOLOGY: DURKHEIM, BOUGLÉ, AND RICHARD

Worms's reviewer, the historian of progress and educational reformer Jules Delvaille, meanwhile represented the new sociological trend in the *Revue* in his contention that national character was less the product of physical and organic characteristics than the result of "conceptions, which are accepted by minds and become principles of individual and collective action."[53] While Ribot had been Tarde's patron and had joined Worms's editorial board, he promoted the career of Durkheim in allowing the publication of *Règles de la méthode sociologique.* Durkheim argued that if race as an organic-psychological phenomenon were truly important, people of different races would not have similar institutions, and people of related races would not have different institutions.[54] While Durkheim's early work in the *Division of Social Labour* had biological models and organic metaphors, thereafter he and his colleagues orchestrated a substantial offensive against the old racial anthropology. He was much less egalitarian, however, on gender issues.[55]

49 Delvaille, Review of Worms, "Philosophie des sciences sociales, t. 1," 661–6, esp. 663.

50 Delvaille, Review of Worms, "Philosophie des sciences sociales, t. 2," 529–31.

51 Delvaille, Review of Worms, "Philosophie des sciences sociales, t. 3," 652.

52 Ibid., 651–5.

53 Delvaille, "La Vie sociale," 583–601, esp. 599.

54 Durkheim, "Règles de la méthode sociologique," *RP* 38 (1894): 27–8; for Durkheimians on physical anthropology, Mucchielli, "Sociologie versus anthropologie raciale," 77–95; on Durkheim see Besnard, ed. *The Sociological Domain* and Lukes, *Emile Durkheim: His Life and Work.*

55 Lehmann, "The Question of Caste in Modern Society," 566–85; Pedersen, "Sexual Politics in Comte and Durkheim," 229–63.

The debate on the nature of social facts became more heated around the turn of the century. The moral and political philosopher Dominique Parodi reported that the vocal and ethically committed Durkheimian Célestin Bouglé did not accept racial purity in history or the importance of ethnic conditions for the triumph of egalitarianism.[56] In the same issue Bouglé himself attacked "biological sociology" by opposing the recent triumph of liberty and equality. Greater liberty and equality were not biological or organic revolutionary tendencies. These tendencies led to constraint and hierarchy. While Le Dantec had seen neo-Lamarckianism as anti-aristocratic and conservatives hailed Weismann, Bouglé accepted Weismann's doubts that acquired traits could be transmitted to descendants. For the badge of alleged superiority was the transmissibility of those characteristics. Cranial deformations were shown to be superficial, limited to the individual. Nor did the scholarly dynasties of Galton and Candolle prove the heredity of talent for science or art, only the transmission of "vague dispositions and indeterminate aptitudes." In a democracy one only needed to equalize the conditions of departure in the struggle for life, which itself could be made more equitable. Natural selection could be humanized in society and in any case sociology was not transposed biology.[57]

In response to Bouglé, the organicist sociologist Alfred Espinas asserted the importance of collective consciousness to the individual. He firmly believed in the biological roots of historical societies, since emotions and desires for the common welfare created the collective consciousness epitomized in the songs and language of the nation. In the aftermath of the Dreyfus affair, Espinas thought the notion of race had been discredited by its abuse, but race could still guide the sociologist in the determination of ethnic families with internal resemblances in their elementary structures.[58]

The Russian émigré Darwinist and organicist sociologist Jacques Novicow also responded that the justified analogies between organisms and societies did not necessarily mean the supremacy of heredity for mental and physical aptitudes. He did not consider that the daughter of Victor Hugo would be poetically talented or the daughter of Thomas Edison would be inventive. At present, the characteristics transmitted by parents were still unknown. Nor did the organicists limit human liberty, since they could easily advocate a very limited non-interventionist state. They argued only that a talented bootmaker had no right to be a ship's

56 Parodi, Review of Bouglé, "Les Idées égalitaires," 545–52; Policar, "Science et démocratie," 86–101.

57 Bouglé, "La Sociologie biologique et le régime des castes," esp. 342, 348–9, 352.

58 Espinas, "Etre ou ne pas être," 453, 471–2, 474.

captain. Castes were a social pathology, but each person should follow his aptitude to exercise a trade. There would be no heredity of professions, but full respect for liberty and equality of rights.[59]

Bouglé's reply to Espinas and Novicow insisted that organicism was unfavourable to democracy since it offered not freedom, but imprisonment in a predetermined aptitude. Organic metaphors also favoured the assumption that the elite are the brains of society. Espinas had not proven the importance of physiology for sociology. Race as the organic basis of a nation was chimerical among humans. Bouglé appealed to Manouvrier's articles on aptitudes and acts (1890) and his refutation of the cephalic index (1899) to argue that only general dispositions are inherited, not professions. Nor was physical transmission of the soul of a nation probable. The flourishing state of Durkheim's periodical, the *Année sociologique*, proved that sociology did not need the help of biology.[60]

Most sociologists, including the Durkheimian Gaston Richard, joined this defence of the primacy of the social milieu. He had argued against Letourneau that the influence of race decreased with civilization. In 1903 Richard strongly attacked both the racial theorist Arthur de Gobineau and Le Bon. The "social fact," he maintained, "triumphs over the anthropological fact and transforms it."[61] Richard insisted that history was about collective character, a product of psychological conditions and the geographical milieu, though among civilized peoples climate was a less significant factor. As Fouillée and Durkheim maintained, history was not just about race and heredity, since the character of peoples changed over time.[62] Richard thought that interpretations of Social Darwinism that trumpeted inevitable racial struggle were false and manifested the bankruptcy of calling race the major determinant of societies.[63]

THE ANTI-DURKHEIMIAN BACKLASH

In 1908 the Russian émigré physician Samuel Jankélévitch, who became a Freudian psychoanalyst, reviewed Bouglé's essay on hereditary castes. Jankélévitch still dared to raise the question of whether physical and mental differences in some past age, when Hindu society was first

59 Novicow, "Les Castes et la sociologie biologique," 365–8.

60 Bouglé, "Le Procès de la sociologie biologique," 122, 129, 136–7, 141, 146.

61 Gaston Richard, Review of Ernest Seillière, "Le Comte de Gobineau et l'Aryanisme historique," 433–6.

62 Gaston Richard, "Les Lois de la solidarité morale," 443, 453.

63 Duprat, Review of Richard, "La Sociologie générale et les lois sociologiques," 515.

organized, were correlated to social differences, even though there was no "clear correlation among physical, mental, and social differences" now. Though Durkheim had similar beliefs, the reviewer worried about the "exclusivism of the sociological school."[64] On the other hand he thought Novicow correctly condemned the deduction of facts from an abusive interpretation of Darwinism.[65] Jankélévitch warned against both deriving social from biological facts and against assuming the independence of social facts from biology. Weismann's theories did not necessarily lead to reactionary social and political views merely because sociologists attributed the principal role to the milieu.[66] Hence Jankélévitch and Bouglé had both seen the positive side of Weismann, unlike Ribot, Le Dantec, and the neo-Lamarckians.

But Jankélévitch certainly represented a mild backlash against the sociological imperative. In 1908 he lambasted "primitive" peoples for reflecting little and not inventing and adapting as needed to renew their milieu.[67] Africans, had they lived in Europe, would have enjoyed the material products of Europe, and learned to use them, but would not have become civilized because they "lack the mental apparatus that permits taking advantage of external physical conditions and creating material goods." He thought there were good scientific arguments supporting both equality and inequality of races and sexes.[68]

Jankélévitch's review of the entire literature on heredity in 1913 concluded that predetermined moral principles often influenced the debate. If acquired traits were transmissible, there would not only be a greater influence of milieu on character but also unlimited progress. If not, merely individual traits would make perfecting the species a work of Sisyphus.[69] The neo-Darwinians would in theory worry about retarding natural selection by social assistance that would benefit only a few of the unfit in the current generation. Jankélévitch was convinced by Weismann's argument that hereditary illness was latent in the germ plasm from previous generations and only provoked by external conditions. Opponents argued that negative alterations were not transmitted except in the case of disease that profoundly changed the organism. He also thought the neo-Darwinian position of selection of accidental variations was more logical than

64 Jankélévitch, Review of Bouglé, "Essai sur le régime des castes," 424.

65 Jankélévitch, Review of Novicow, "La Critique du Darwinisme social," 537–41; Paulhan, Review of Novicow, "La Morale de l'intérêt dans les rapports individuels et internationaux" and "Mécanisme et limites de l'association humaine," 290–301.

66 Jankélévitch, "Essai de critique sociologique du Darwinisme," 478, 482–3.

67 Jankélévitch, "Du Rôle des Idées dans l'évolution des sociétés," 256–80.

68 Ibid., 262 (citation), 266, 278.

69 Jankélévitch, "La Position actuelle du problème de l'hérédité," 551.

purposeful adaptation of human beings. Further research was needed on how such reasoning applied to races, peoples, ethnic groups, and social classes.[70] So Jankélévitch was more critical than most contributors of neo-Lamarckianism but skeptical about the complete abandonment of racial theory by the Durkheimians.

The psychologist, philosopher of religion, and criminologist Henri Joly (1839–1925) also viewed with some reservations the journalist Jean Finot's plea to abandon racial prejudice. An optimistic future of racial fusion would be impossible, Joly thought, since Africans could not yet approach European capabilities, though Asians were rapidly catching up. Joly agreed that "Psychic and social habits modified the type, rather than the reverse." But he was also sure that the conquests of the "hardiest and most favoured races" would profit all humanity. He agreed with Finot that races were neither inevitable enemies nor "irreducible" to each other. However, Joly saw an equal prejudice in thinking that differences were artificial and that civilization would make them disappear. The ideal state would be one where each race was distinct enough to avoid deadly competition.[71]

The attack on racial hierarchy by Durkheimians such as Richard and journalists such as Finot represented a new trend in the articles and reviews of the *Revue philosophique*. But some authors showed reluctance to discard this concept. The Dreyfus affair had helped galvanize Durkheimians into defence of democratic values. Other authors of Jewish background, such as Worms and Jankélévitch, though not endorsing simplistic racial theory, would not abandon a biological approach to sociology.

GENDER LIMITATIONS

There was no clear-cut Durkheimian offensive on gender issues, since even if authors did not believe in an essential nature of women, they wished for a social order that envisaged women as tending the household and caring for children. Some believed the family would decline if women were given full equality of opportunity and legal rights. For many authors the topic of gender offered less potential flexibility than race. Unlike the Société de sociologie, which allowed women to lead some discussions, the *Revue* featured very few works by women authors (an exception was the anthropologist Clémence Royer, but she did not there articulate her feminism).

70 Ibid., 552–3, 559–60, 566–7.

71 Henri Joly, Review of Jean Finot, "Le Préjugé des races," 101, 104 (citation), 107.

The *Revue* had analyzed both the essentialist Le Bon and the sociological anthropologist Letourneau, who finally arrived at a strong feminist perspective. Richard reviewed Letourneau's posthumous publication assessing women's condition in diverse races and civilizations.[72] Once again, Letourneau highlighted an approach to social phenomena as the unequal progress of distinct races toward complex societies. But in his conclusion, Letourneau could not approve the injustice of women's deteriorating situation in complex societies. He now argued against an essential organic nature of women that constructed a "feminine type." Rather, social conditions in civilized societies resulted in division of labour that deprived women of intellectual culture and insulated them from an active and free life.[73] Given this conclusion, Richard saw little point for Letourneau to advocate an evolutionist and naturalistic philosophy that could by implication approve the worsening situation of women and would be at variance with his feminist values.

Richard's own survey of women in history credited women with helping to civilize men by moralizing them. A democratic civilization, he thought, would assimilate the sexes, but would conserve the "social cult of the maternal function."[74] However, earlier Richard had still believed there was feminine intellectual regression during menstruation and pregnancy. He rejected overly hasty conclusions that smaller brain size and reproductive functions harmed intellectual potential. But he insisted that a radical member of the Paris Société de sociologie Jacques Lourbet, who thought social restrictions hampered the fulfillment of women, had not set sufficient limits to liberty for women. There was still an eternal feminine nature.[75]

Alfred Fouillée's study of French temperament and character commented that feminine intelligence was less innovative and reflective, but more morally sensitive – gentle, calm, and affectionate. The present feminist movement correctly based its claims on the merit of equal duties and rights within the family and civil society. But a drive for "perfect equality" misunderstood the radical difference in the character and aptitude of the sexes.[76] Like Manouvrier, Fouillée feared that if women abandoned the family, marriage would be devalued, love would become mere sensual satisfaction, prostitution would be rampant, and there would be

72 Gaston Richard, Review of Letourneau, "La Condition de la femme," 432–3.

73 Letourneau, *La Condition de la Femme*, 501–7.

74 Delvaille, Review of Richard, "La Femme dans l'histoire," 416–17.

75 Gaston Richard, Review of Lourbet, "La Femme devant la science contemporaine," 434–5.

76 Souriau, Review of Fouillée, "Tempérament et caractère," 629–30.

more abortions, infanticide, and abandonment of children. Hence, the future of the race was at stake.

Paul Lapie, a Durkheimian sociologist and author of works on moral philosophy and pedagogy, proclaimed equality of the sexes and approval of legalized divorce, but insisted that men needed affection and calm in the family.[77] Jacques Novicow, an ardent supporter of free union between the sexes without legal marriage, judged that feminist claims were quite legitimate and that the absolute juridical equality of man and woman fulfilled the perfect condition of the species.[78] An index of civilization was how well a country fulfilled this ideal. Muslim and Oriental men were deprived because they did not educate their women. With some exceptions, such as Novicow and the latter-day Letourneau, the social imperatives of the household and family trumped the more radical feminist aims for contributors to the *Revue philosophique*.[79]

CONCLUSION

The debate over heredity and milieu was characteristically complicated by the different powers ascribed to the milieu in different eras. The hereditarian view never claimed that milieu had been a nullity, especially during racial differentiation, but only that race was now predominant through transmission of what was fixed in the organism. Neo-Lamarckian anthropologists, even when most enthusiastic about racial hierarchy in principle, allowed for inheritance of acquired characteristics. But they applied an invidious distinction to non-Europeans that tempered their capacity for progress.

Taine and Spencer had allowed for the influence of both race and milieu, but their critics and disciples, with some exceptions, tended to emphasize the perceived principle of racial stability. The anthropological countercurrent began to doubt the importance of the physical attributes measured in anthropometry and to suggest that milieu and other social factors might indeed allow for substantial improvement.

The ambivalence of theories about heredity itself had also become evident in the varying responses to Weismann and neo-Lamarckianism. Republicans eager for progressive social change had endorsed neo-Lamarckianism, but authors like Ribot had used neo-Lamarckianism to insist on the fixing of acquired traits by hereditary transmission. Bouglé felt that Weismann seemed more conducive to democracy, while Le Dantec thought that Weismann had the potential to justify aristocracy.

77 Duprat, Review of Lapie, "La Femme dans la Famille," 664.
78 Novicow, "Les Bases biologiques, psychologiques et sociologiques du droit," 7.
79 See also Surkis, *Sexing the Citizen*.

The great philosophers and sociologists such as Fouillée, Tarde, and Worms would cling to the significance of race but not to its predominance. Sociologists such as Durkheim, Bouglé, and Richard would emphasize social factors, while philosophers such as Joly and physicians such as Jankélévitch were afraid to be swept too far by this trend.

No doubt the ferment over the "new woman" and the movement for Civil Code reform and political equality for women led to reconsideration of the principle of an essential nature of women.[80] Extremist anthropologists such as Le Bon subscribed to this principle. As we shall find in the Société de sociologie, a desire to keep traditional family and social roles was a critical factor working against wholesale emancipation of women for thinkers such as Fouillée, Richard, and Manouvrier. Social order had to be protected, even if nature were not indelible. However, Letourneau and Novicow were exceptional in anticipating an egalitarian future. Liberalism on race and liberalism on gender were not necessarily correlated, with Letourneau showing a dissonance on the two issues while Novicow was liberal on both.

The variety of viewpoints even on such a charged issue as psychological heredity shows that the editor Ribot did not tyrannically impose his own views on authors of articles and reviewers. The prewar years before 1914 had witnessed the decline of the prestige of the "master thinkers" of the 1870s, Taine and Spencer. The Dreyfus affair had consolidated the identity of Durkheimian sociologists and induced doubts about the old racial theories so dear to Le Bon and conventional anthropologists. Nevertheless, the incipient backlash and the persistence of Le Bon's influence held forebodings of a revival of racial theories.

The controversy about the relative proportions of the influence of heredity and milieu could create disputes that traversed the emerging social sciences and could have significant political and social consequences. While hereditarians could be expected generally to advocate racial hierarchy and an essential nature of women, the complexity of the discussion in this major philosophical periodical would show fluid and surprising alignments of views. Belief in the supremacy of milieu, whether on questions of race or gender, may have been necessary, but not sufficient, to move authors in the direction of egalitarianism. It would take the major cultural upheaval of the period after World War II and the era of decolonization to uproot the racial ideology endorsed by the old physical anthropology and to overcome the cultural inhibitions that prevented full citizenship for women.

80 McMillan, *France and Women, 1789–1914*; Klejman and Rochefort, eds, *L'égalité en marche*; Roberts, *Disruptive Acts*.

6

Alfred Binet: from Measuring Heads to Testing Intelligence

THE FAMOUS INVENTOR of the "metric scale of intelligence," so influential on later IQ testing, Alfred Binet (1857–1911), benefited from the patronage of Ribot, who published his early papers in the *Revue philosophique.*[1] Unlike Ribot, he was unsuccessful in obtaining an appointment in philosophy at the Sorbonne, in part because he had a Faculty of Sciences rather than a Faculty of Letters doctorate. However, in later years the Société française de philosophie had no qualms about accepting him as a lecturer.[2] By the late 1870s he experienced the same intellectual influences as Ribot – Taine's quest for scientific psychology, the associationism of John Stuart Mill and Alexander Bain, Galton's search for individual differences, and the evolutionism of Herbert Spencer. He became an assistant in Charcot's hypnosis and hysteria lab at the Salpêtrière and for a long time shared Ribot's interests in suggestion and hypnosis. Binet joined Charcot's Société de psychologie physiologique in 1885. He also valued Taine's critique of metaphysics and concern for exceptional psychological phenomena (prodigies or multiple personalities) over Ribot's evolutionism.[3]

At least by 1888 Binet began to think about problems of generation, if not heredity, when he summarized for the *Revue philosophique* the latest conclusions about germ plasm and embryology from his father-in-law Etienne Balbiani's course at the Collège de France on reproduction and heredity.[4] But while Ribot continued to write about pathological

1 Mucchielli, "Aux Origines de la psychologie universitaire," 281; Wolf, *Alfred Binet*, 40–3.

2 Archives nationales, Procès-verbaux de l'Assemblée des professeurs, AJ16 4749; Elections to Collège de France, F17 13356; Nicolas and Ferrand, "Alfred Binet and Higher Education," 264–83.

3 Plas, "La Psychologie pathologique d'Alfred Binet," 234.

4 Binet, "Résumé des conférences de Balbiani," 529–59.

psychology, Binet moved fitfully toward normal psychology after 1890. He began with the observation of his own daughters, Madeleine, aged five, and Alice, aged three, for their motor abilities, reaction times, and perception of size.[5] A chance meeting with the Nancy psychologist Henry-Etienne Beaunis (1830–1921) in 1891 led to Binet's employment (without compensation) as a major figure in the new psychology lab linked to the Sorbonne and nominally under the jurisdiction of the Ecole pratique. By 1894 he would direct the lab and launch a major series of comprehensive investigations into individual development. In the same year he founded the most important periodical for the development of normal psychology at the time, *L'Année psychologique*, modelled after the philosophical yearbooks with original articles, surveys of the literature, and book reviews.[6] In his survey of experimental psychology for students and practitioners, Binet proved to be eclectic in his consideration of proper methods – from introspection to questionnaires to study of the customs of various peoples.[7] Even though Ribot himself was not an experimentalist, his letters to the psychologist Henri Piéron indicate some veiled skepticism about the value of Binet's work in the *Année psychologique*. He did not highly value Binet's questionnaires about the teaching of philosophy and claimed that the 1908 volume of the *Année psychologique* was very diffuse and difficult to review.[8]

BINET AND HEAD MEASUREMENT

From 1895 Binet's primary interest was a program of comparative individual psychology. In the second volume of *L'Année psychologique*, Binet published with his collaborator Victor Henri a manifesto on individual psychology that was a critique of previous testing procedures and a program to investigate variations in memory, association of images with words, attention, the faculty of understanding, suggestibility, esthetic and moral sentiments, muscular force, will, and ability to grasp concepts.[9] Shut out of secondary schools by obstructive administrators, he instead observed his daughters and primary school pupils with experiments on muscle physiology, blood circulation, and both physical and mental endurance.

5 Wolf, *Binet*, 80–2; Fancher, *The Intelligence Men*, 58.

6 Nicolas, "Alfred Binet et 'l'Année psychologique'," 665–99; Nicolas, Ségui, and Ferrand, "Les premières revues de psychologie," 71–110; Carpintero and Molto, "*L'Année psychologique* et la naissance de la psychologie expérimentale," 15–34.

7 Binet, *Introduction à la psychologie expérimentale*, 145–6.

8 Carroy and Plas, "La Psychologie: Science naturelle et science morale?" 346–8.

9 Binet and Henri, "La psychologie individuelle," 411–65; Bornstein, "Mesure de l'intelligence et pédagogie," 148–72.

During most of these studies, Binet remained non-committal on causes of variation in abilities. But he was confident that there were different classifiable character types. Like Galton, who had tested differences in sensitivity in the hopes of relating them to intelligence, and Lombroso, who had sought stigmata of criminals, the authors believed it important to classify empirically the characters of children as well as of adults such as criminals. By 1896 he had done several experiments with school children viewing pictures, which led to classification into types of observers: those who observe a whole picture at once: those who are good at describing the picture in detail; those who express strong emotions without attention to detail; and those who remember pictures from previous encounters, but don't tell what they see.[10]

Binet illustrates the interest in anatomical or physiological correlations with mental traits. In the end, he would turn away from a hereditarian interpretation without committing himself to a definitive assessment of hereditary versus environmental factors. But he represents the emergence from the anthropological shadow toward a more pronounced emphasis on surrounding sociological factors.

Binet wished to investigate the typical anthropological assumption that physical dimensions of head and body and physiological processes correlate with mental capacities. His interest in anthropology is even observable in his labelling of the sensory psychological studies of the American psychologist Joseph Jastrow as "mental anthropometry."[11] Tellingly, when a friend, the physiologist and photographer Etienne-Jules Marey, recommended Binet for the Collège de France in 1902, he praised Binet's "anthropological approach."[12] Binet hoped that anthropologists would assist his ethnic, pedagogical, medical, and judicial studies.[13] Even after the Durkheimians and Dreyfusard intellectuals assaulted the primacy of heredity and warned of the dangers of racism, Binet persisted with physical measurements. There is no evidence that Binet himself was a racist, however, and he privately endorsed the "beautiful cause" of Dreyfus.[14]

10 Binet, "Psychologie individuelle: la description d'un objet," 296–332, esp. 318–23; Garanderie, preface to Alfred Binet, *Œuvres completes, t. 20, L'Etude expérimentale de l'intelligence*, xxi.

11 Undated manuscript in Archives nationales, Papiers de Binet, 520 AP 44.

12 Archives nationales F 17 13356, Assemblée des professeurs du Collège de France, 19 janvier 1902, Rapport de Marey.

13 Binet and Henri, "La Psychologie individuelle, Revue générale," *AP* 2 (1896): 434; Bornstein, "Mesure de l'intelligence"; Andrieu, "Céphalométrie scolaire et intelligence inégale," 157–74.

14 Bibliothèque nationale, Manuscrits, N. A. F. 24488, fol. 55, Binet, Lettre à Louis Havet, 19 août 1899.

In 1897–98 Binet, with his Romanian student collaborator Nicolas Vaschide, began a long series of psychophysiological observations – heart rate, circulation, respiration rate with the spirometer, muscular strength with the dynamometer – that were correlated with each other and with signs of intelligence as determined by teachers.[15] The properly anthropological protocol of measuring head size began in 1898 with Vaschide and the young alienist Théodore Simon, of the Perray-Vaucluse institution for mentally challenged children and adolescents at Epinay.[16] Binet retained the hope that he would be able to find some correlation, although by 1908–9 his enthusiasm had flagged considerably.

A pivotal figure in Binet's fascination with anthropometry was the skilled but skeptical assessor of head dimensions, Léonce Manouvrier. He represented at once the refusal of anthropologists to give up measurement and the recognition of their possible futility.[17] The historical account attributed to Binet and Vaschide directly cited Manouvrier's article on brain structure (1897) for Charles Richet's *Dictionnaire de physiologie.*[18] There Manouvrier argued, "qualitative superiority of the brain was a condition for intellectual superiority, but quantitative superiority was another condition, and morphological superiority was still another condition."[19] Cell structure, chemical composition, nutrition, and energy of temperament might all be factors in intelligence, but so were "conditions of education, family milieu, social milieu, civilization, and language." Still, no other condition could replace "superiority of weight of the brain," which had a "value all its own of a special order."[20] Unsurprisingly, then, while Manouvrier elsewhere highlighted social factors, his unsuccessful application for a philosophy chair at the Collège de France in 1899 still asserted, "anatomical variations existing between the races, sexes, ages, and individuals" might have broad significance, including "more or less direct relation with intellectual and moral character."[21]

When Binet undertook his own program of head measurement, his historical essay quoted in full Broca's memoir showing that education

15 Binet, "La Mesure en psychologie individuelle," 113–23, esp. 123; also Binet and Vaschide, "Corrélation des épreuves physiques," 142–72, and "Corrélation des tests de force physiques," 236–44; comments by J. Larguier de Bancels, "L'œuvre d'Alfred Binet." 151–66.

16 [Binet and Vaschide], "Historique des recherches sur la céphalométrie," 245–98.

17 Blanckaert, "La Crise de l'anthropométrie," 157–8.

18 [Binet and Vaschide], "Historique des recherches sur la céphalométrie," 296–8; Manouvrier, "Cerveau," 670–745.

19 Ibid., 699; cf. Manouvrier, "Les Qualités intellectuelles," 65–84.

20 Manouvrier, "Cerveau," 712, 714–15.

21 Archives du Collège de France G-IV-f 1897–1900, Manouvrier, "Considérations présentées par M. L. Manouvrier à l'appui de sa candidature, " fol. 2.

itself could increase brain volume. He also cited Galton and John Venn on the greater cranial capacity of superior Cambridge students.[22] Though somewhat contemptuous of the methodology of the German anthroposociologist Otto Ammon, a disciple of Vacher de Lapouge, Binet still thought him worth quoting on the alleged relationship of intelligence to cephalic index.[23] From his historical research, Binet would assert as an unassailable proposition that "intelligence and head volume, as evaluated by approximate measures on the living subject, have a real relationship verified by all methodical investigations without exception."[24] The inaccessibility of the brain itself might not matter since the frontal lobe seemed to be the same percentage of the total cerebral weight no matter what the brain size. While head size was not proportional to intelligence in linear fashion, it appeared to be a legitimate indicator.[25]

Binet's collaborator, the alienist Théodore Simon, wrote a medical thesis that attempted to resolve the contradictions in the literature between Galton, who, apart from head size, found no correlation between general physical development and intelligence, and the positive correlations of the American W. Townshend Porter.[26] Simon showed that in almost every physical indicator, including head size, which he measured by circumference only, the "idiots, imbeciles, and retarded" population lagged behind the "normal" values for the various age groups.[27]

In 1901 Binet refined his measuring techniques to follow the procedures of the anthropologists Broca and Topinard: measurement of anterior-posterior diameter of the head, transversal diameter of the head, and distance from chin to the bottom of the nose. Manouvrier provided elaborate instructions and assisted, along with the physiologist Louis Lapicque and Muséum librarian and anthropologist Joseph Deniker, in performing the actual measurements in schools and institutions in 1900–1901. Deniker later measured residents of the Institute for the Deaf-Mute, while physicians or alienists measured inmates of the asylums.[28]

A new round of measurement in the Paris primary schools gave "surprisingly insignificant results," with only a millimetre difference between

22 [Binet and Vaschide], "Historique des recherches de la céphalométrie," on phrenology and Broca, 246–9; Wolf, *Alfred Binet*, 153–60.

23 [Binet and Vaschide], "Historique," 288–9.

24 Ibid., 294; cf. Andrieu, "Céphalométrie," 161; Gould, *The Mismeasure of Man* (1996), 176–7.

25 [Binet and Vaschide], "Historique," 256, 295–6; Andrieu, "Céphalométrie," 162.

26 Simon, *Documents relatifs à la correlation*, 7.

27 Simon, *Documents*, 15; Chapuis, "Binet, une passion de la mesure," 71–81.

28 [Binet and Vaschide], "Historique"; Binet, "Etudes préliminaires de céphalométrie," 369–74; Binet, "Recherche sur la technique de la mensuration," 320–1, 325, 363; Chapuis, "Binet, une passion de la mesure," 77.

the "intelligent" and the "unintelligent." Hence, Stephen Jay Gould portrays Binet as soon losing patience with the tenuous physical-mental correlation.[29] However, more rigorous samplings showed Binet that for most measures he could arrive at a sensible difference of 3 to 4 millimetres between "elite" and "average" pupils, though the "retarded" clearly did not follow a rule of proportionality, since there were some very large and some very small heads.[30]

Contrary to Gould's interpretation and to Andrieu's presentation of Binet's increasing skepticism on anthropometry, Chapuis has argued convincingly that even until 1910 Binet relentlessly desired to find indicators of intelligence, whether head size or another factor. Unlike Ribot, whose interest in anthropology stemmed from his Spencerian evolutionism and hereditarianism, Binet's fascination with measurement was responsible for his connection with anthropology.[31]

In 1902, while recording the contradictory positive correlations reported by the Czech researcher H. Matiegka, Binet conceded that Galton's disciple Karl Pearson had found no relation between cranial capacity and the most common appreciations of intellectual capacity. At the same time Binet was still doggedly trying to establish benchmarks for normal growth of the face and cranium, as well as other physical indicators such as height, weight, muscular strength, and pulmonary capacity, among a relatively homogeneous population of schoolchildren.[32]

An interesting sidebar to these studies is the continued interest of Binet's former collaborator Nicolas Vaschide, with the co-operation of socialist feminist Madeleine Pelletier, in physical indicators of intelligence. Using her medical training before embarking on a psychiatric career, Pelletier, under the tutelage of anthropologists Manouvrier and Papillault, helped Vaschide conduct a study from 1901 to 1903 that accepted the basic anthropological paradigm of connecting head size to intelligence. They also argued that intelligence was mostly innate and that most humans "could not surpass mediocrity despite the most careful education and the most favorable milieu." In addition, they accepted the evolutionary superiority of Europeans, compared to "inferior" races.

29 Binet, "Recherches de céphalométrie sur 26 enfants d'élite," 403; Gould, *Mismeasure* (1996), 180–3; Lesonn, "Les déboires d'enquête d'Alfred Binet," Retrieved July 31, 2007 from http://perso.orange.fr/koebel/afs/resumesL.htm#T08

30 Binet, "Recherches de céphalométrie sur des enfants d'élite (déc. 1900 et janvier 1901)," 422–4, 427–8.

31 Andrieu, "Céphalométrie," 166–7; Chapuis, "Binet, une passion de la mesure," 72–3.

32 Binet, "Nouvelles recherches de céphalométrie," 341, 344; Binet, "La Croissance du crâne et du face," 345–88.

While Pelletier would break with anthropology by 1904 as her feminism crystallized, the incident serves to show that there was no automatic emancipatory front for the liberation both of women and of non-Europeans.[33] Karl Pearson used the flaws in the study of Vaschide and Pelletier to argue against a simplistic correlation of head size and intelligence.[34]

BINET AND INTELLIGENCE MEASUREMENT

In 1904, the bulletin, edited by Binet, of the pedagogical study group founded by Ferdinand Buisson in 1899, the Société libre pour l'étude psychologique de l'enfant (Ribot was the first honorary president) featured an article in which Binet compared various head measurements of "normal children" and "idiots." The society established a commission on "abnormal" children that led directly to the government commission of the Public Instruction Ministry of December 1904, with the aim of setting conditions for admission to special education classes (medical or pedagogical tests were deemed most important). Binet's purpose in designing his famous metric scale of intelligence was to develop, before the conclusion of the work of the commission, an additional psychological test to identify and assist those slow learners who could benefit from schooling rather than institutionalization.[35] Despite common present-day critiques that Binet acquiesced in marginalization of the disadvantaged, he indicated his hope that the "retarded" would achieve their potential through special classes and mental "orthopedic" exercises. Binet was not condemning them to stagnate with a fixed measurement of intelligence (such as the IQ used later by other psychologists). His definition of retardation was a relative classification of two years behind grade level below age nine and three years behind above age nine.[36]

However, Binet was undoubtedly a partisan of social stability. In assessing the need to promote the social utility of slow learners, he expressed fears that apathy toward fellow citizens was a recipe for violent

33 Sowerwine, "Woman's Brain, Man's Brain," 299–300; the full text of Vaschide and Pelletier was published as N[icholas] Vaschide et Madelaine (sic) Pelletier, *Recherches Expérimentales sur les Signes Physiques de l'Intelligence*; Gordon, "Convergence and Conflict," 141–63 points out Pelletier's endorsement of Manouvrier's refusal to admit a cerebral basis of inferiority of women and her later conversion to Letourneau's neo-Lamarckianism and abandonment of the primacy of heredity.

34 Lee, Lewenz, and Pearson, "On the Correlation of the Mental and Physical Characters in Man, Part II," 107–8.

35 Nicolas, "Alfred Binet 1857–1911," 15–28; Huteau, "La Réception de l'Echelle Métrique," 41–72.

36 Dufour, "Binet et la pédagogie," 163–72; Mougniotte, "Alfred Binet et 'la chose politique'," 41–9; Wolf, *Binet*, 194.

revolution.[37] He also thought the practical application of psychology would counteract the "hysterical susceptibility" of crowds and would help to "discipline, while giving representation to, professional interests" (so evident in the recent violent postal strike).[38] Binet worked in a political context in which Third Republic "solidarism," as expressed through the educator Buisson and the government minister Léon Bourgeois, chair of the interministerial committee on abnormal children, would favour containing revolutionary tendencies while expressing obligations to social co-operation.[39]

Binet did believe his tests revealed a "natural intelligence" different from school achievement that itself was a product of attention, regular habits, and sustained effort. He felt that arranging weights in order or defining abstractions and interpreting pictures would be less subject to variations in the milieu and provide a more adequate explanation of spontaneous intelligence. He ignored the possible bias of a legion of culturally sensitive questions, such as estimates of numbers of coins, knowledge of the price of commodities, and moral assumptions in moral dilemmas.[40]

Whatever his overconfidence in the cultural and social neutrality of his test, he moved beyond Ribot and most anthropologists (with conspicuous exceptions such as Manouvrier) in highlighting the direct relationship of social circumstances to physical development and then to intelligence. This awareness does not seem to be a product of either proximity to the Durkheimian sociological school or overt political engagement. However, the Société libre had an obvious interest in the social context of pedagogy and considered education itself as a potential corrective to the social problems of poverty.[41]

ANTHROPOLOGICAL AND SOCIOLOGICAL APPROACHES

Even while conscious of the shortcomings of anthropometry, Binet concurrently sought physical indicators of intelligence.[42] To avoid the problem of misleading mean values, he developed the notion of "anthropological boundaries." A "frontier" or boundary was a value beyond which most measurements were either of "normal" or "abnormal"

37 Binet and Théodore Simon, *Les Enfants anormaux*, 2.

38 Binet, "Le Bilan de la psychologie en 1908," v-xii.

39 Foschi and Ciccioli, "Politics and Naturalism," 267–89.

40 Binet and Simon, "Le Développement de l'intelligence chez les enfants," 74–5, 78.

41 Binet, "Ce que vaut l'école primaire," 58–63; Apfelbaum, "Origines de la psychologie sociale en France," 404–5; Foschi and Ciccioli, "Politics and Naturalism," 275.

42 Cf. Andrieu, "Céphalométrie," 170.

children. He still attempted to correlate intelligence with anatomical and physiological measures as well as with physiognomy and stigmata. For example, a 14-year-old boy only 1.35 m (4'5") tall would have only a one-in-twelve chance of having normal intelligence, and a twelve to one chance that he was "retarded."[43] Similarly, for the sum of anterior-posterior diameter of the head and width of head, "normals" averaged 20 mm more than "abnormals." Thirty-four per cent of "imbeciles" were under the average for at least one dimension of head measurement. He acknowledged though that similar physical measurements do not guarantee similarity in sentiments and character.

In 1905, in his study of girls aged 6 to 14 in primary and middle schools, the more intelligent had a larger anterior-posterior diameter – "experiments of cephalometry confirm those of memory."[44] With V. Vaney, the principal of the rue La Grange-aux-Belles elementary school in the Belleville section of Paris that became a veritable laboratory for Binet, he encouraged investigation of relationships between medical measures, height, weight, and head measurement as possible signs of retardation, though these would be less frequently applicable than the metric scale of intelligence.[45] He also hoped eventually for comparisons based on sex, social conditions, race, and intellectual status, taking account of the differences between normal and abnormal children.[46]

Binet's ultimate verdict on head measurement was that the intelligent had larger heads, but the correlation was "very weak" and not good for individual diagnosis. He concluded that the "abnormals" had both microcephalics and macrocephalics among them. While "normal" children were rarely under the "anthropological boundary" for height, and only about one-sixth were below the frontier for anterior-posterior diameter, one-third to 40 per cent of the more troubled inmates of Bicêtre asylum were below the boundaries for height and head measurements.[47]

Binet also remained fascinated with contrasts in physiognomy, where facial views seemed more revealing than profiles. A photography experiment conducted by a Paris primary schoolteacher, Mme Rousson, noted that for the least retarded, physiognomy seemed most useful. Binet added that it would be helpful if these signs "were known by all and could be used by all."[48] Teachers were more astute in reading this physiognomy

43 Binet, "Les frontières anthropométriques des anormaux," 433, 438.
44 Binet, "À propos de la mesure de l'intelligence," 82.
45 Binet and Simon, "Méthodes nouvelles," 244.
46 Binet and Simon, "Application des méthodes nouvelles," *AP* 11 (1905): 246.
47 Ibid., 245–346; Binet and Simon, "Méthodes nouvelles," 236–8.
48 Binet, "Note sur l'étude de Mme. Rousson," 702.

compared to outside observers, enjoying a success rate between 63 and 79 per cent in placing students accurately compared to 16–41 per cent for others. The next year, in attempting to sort identification-style photographs taken according to the techniques of demographer and Paris police statistical director Jacques Bertillon, both men and women teachers gave correct answers averaging 77–79 per cent for "normal" and "abnormal" children.[49] Even while Binet was losing confidence in the cephalometric standards, he hoped to return to this promising technique.[50]

Handwriting experts were also successful in distinguishing age, sex, and intelligence of pupils better than chance – though not the character of a murderer. Binet believed this graphology study, which occupied one of the committees of the Société libre, seemed to be useless for character, but, if pupils were classed by age, experts could recognize intelligence through their handwriting.[51]

We cannot determine whether Binet knew of Manouvrier's attack in 1911 (the year of Binet's death) on precise anthropometric measures for aptitudes in sports, the professions, or the military service. Manouvrier concluded that many attributes of intelligence, such as memory, assimilative capacities, and linguistic facility, did not depend on brain mass alone. He even implicitly extended his attack on psychological aptitude scales, since he thought the "insufficiency of physiological analysis showed that psychological analysis would yield analogous illusions."[52]

By 1910 Binet had clearly lost most of his confidence in physical indicators as a stand-alone method: "Nothing can be concluded about a child's intelligence from corporal anthropometric measurements."[53] So-called stigmata (blemishes or deformities of the body, as Lombroso thought characteristic of many criminals) were not measurable. They were not necessarily indicative of intellectual deficiency, since malformation can befall anyone. One "needed to judge by intelligence, not the form of the head, eyes, and ears." Even now, though, he refused to rule out physical indicators, since a critical usage of them was necessary, and no aid to discerning intelligence should be summarily rejected.[54] The "consideration of physical signs of intelligence can add a little confirmation and an explanation

49 Binet, "Lecture d'une physionomie d'enfant," 176; see also Binet, Simon, and Vaney, "Recherches de pédagogie scientifique," 233–74.

50 Binet and Simon, "Méthodes nouvelles," 241–2.

51 Binet, Simon and Vaney, "Recherches de pédagogie scientifique," 270.

52 Manouvrier, "Anthropométrie et aptitudes," 481–83, 486.

53 Binet, "Les Signes physiques de l'intelligence chez les enfants," 29–30; Binet, *Modern Ideas about Children*, 52.

54 Nicolas, "Le Binet-Simon et les origines de la mesure de l'intelligence," 24–5.

to a judgment made for other reasons, and according to other documents; because the intelligence of a child is demonstrated and proven uniquely by intellectual manifestations."[55] The truth about physical correlations with intelligence is "neither with the skeptics nor the credulous." Indeed "cephalometry confirms some aspects of phrenology."[56] Normal pupils had a 50 per cent chance of being more intelligent, with larger or longer cephalometric measures, and an 80 per cent chance of not being retarded. Similarly, the retarded were more likely to have lower than mean values of cephalometric measures.[57] But only the tendencies were significant. Binet also remained undecided on the primacy of physiology or psychology – though he admitted that circulation, nerve cells, and cell chemistry all might play a role.[58]

Similarly, Binet's tests did not discern whether problems with intelligence were organic and did not distinguish between "acquired and congenital idiocy."[59] In the 1911 revision of the metric scale of intelligence, he admitted that both intelligence and retardation might be hereditary in some form, but he could not determine how.[60] He saw the intertwining of social and hereditary organic factors in the fact that the poorer classes are shorter, weigh less, and have smaller heads and less developed muscles.[61]

Earlier, he had noted that possible hereditary influences on retardation might be the age of parents, the consequences of alcoholism (afflicting 42 per cent of the fathers of those with poorest intelligence) and the legacy of tuberculosis (13 per cent of fathers of the same group), epilepsy, or hysteria (18 to 25 per cent of fathers showed nervous system symptoms). However, he also thought the lastborn in families might have difficulty because their families could not adequately provide for them.[62]

For all his fascination with the physical and anthropometric, then, Binet recognized the direct relationship of social circumstances to physical development. As a measure of poverty, he used the criterion of whether schools were donating free canteen meals and various amounts of clothing to the pupils (the amount of clothing distinguishing the poor from the abjectly poor). In addition, he classified the profession of parents and the number of children in the family to set off the abject

55 Binet, "Les Signes physiques," 14–16; Binet, *Modern Ideas,* 53–4.
56 Binet, "Les Signes physiques," 5–6.
57 Binet, *Idées modernes,* 57, 59.
58 Binet, *L'Etude expérimentale de l'intelligence,* 305.
59 Binet and Simon, "Méthodes nouvelles," 191.
60 Binet, "Nouvelles recherches sur la mesure du niveau intellectuel," 173.
61 Binet, "Nouvelles recherches sur la mesure du niveau intellectuel," 190.
62 Binet and Simon, "Application des méthodes nouvelles," 233–4.

and the poor from the middle income (parents often manual workers or office employees) and comfortable groups (either two working-class parents or a parent in commerce and industry). He then measured about one hundred boys and one hundred girls for the breadth of shoulders, cranium and face size, including transversal and biauricular diameters, and various distances in the face.[63] Using these divisions, he found that two-thirds of the physically retarded were from the two impoverished groupings.[64]

The lower socioeconomic classes obviously suffered from poor nutrition and in some cases from poor hygiene. In addition, advanced physical development correlated weakly with advanced intellectual development, as did retarded physical development with retarded intellectual development. In the *Bulletin*, he summarized his findings as showing that there was a one-sixth chance that intellectual development would be the reverse of physical and a one-third chance that retarded body and mind would go together or that an advanced mind and advanced body would go together.[65] The obvious conclusion was that healthy functioning of physical organs and, where appropriate, free canteen food and free clothing, would make a difference to the intellectual development of impoverished children.[66]

In 1909 he reiterated that upper classes had continual opportunities to learn, while "misery and poverty not only lower physiological measures but result in the miserable attrition of intelligence among primary school children." The absence of parental affection could add to the effects of poverty itself. Sociologists admitted that "social classes continue to exist" and their effects were "equally confirmed by physiology, psychology, and morals."[67] Poor nutrition and poor hygiene could be an expression of social misery affecting facial features, postures, and physiology. Alcohol could lead to physical degeneration that in turn promoted intellectual and moral degeneration.[68]

From the outset he admitted that race would affect height measurements, and that one could not judge a group of Japanese subjects retarded if they were below the mean for French. But in the schools

63 Binet, Simon, and Vaney, "La Misère physiologique et la misère sociale," 1–24; see Chapuis, "Binet," 72–3 for a list of articles discussing Binet's cephalometry and a corrective to Gould's claim that Binet moved away from these techniques by 1910.

64 Binet and Simon, "La Misère physiologique."

65 Binet, Simon, and Vaney, "Le Laboratoire de la Rue Grange aux Belles," 10–24.

66 Binet, Simon, and Vaney, "Recherches de pédagogie scientifique," 233–74.

67 Binet, "Ce que vaut l'école primaire," 58–63, and *Bulletin de la Société libre pour l'étude de la psychologie de l'enfant* 54 (1909): 101–8, cited in Wolf, *Binet*, 312.

68 Binet, *Idées modernes*, 61–2, 64, 66.

examined, he insisted there was no significant ethnic deviation.[69] Never overly fascinated with race as a variable, he still in 1911 hoped to apply a comparative intelligence test to children of French, Annamite (Vietnamese), and African (Dahomey) descent.[70]

SOCIAL IMPLICATIONS

Despite Binet's sympathy for the plight of the poor, he suggested that it would be good if "each worked according to his recognized aptitudes, in a way that no parcel of psychic force would be lost to society." Especially useful would be the identification of aptitudes for those of inferior degrees of intelligence who might be helped. This reflection sounds akin to the phrenologists' desire to slot each personality type into the right social niche.[71] In 1909 he thought this the "most important matter in education and teaching; one should educate them according to their aptitude and also orient them to a profession."[72] He recognized that there would be some with practical rather than literary skills, and that they would go to professional and technical schools, where they would develop manual skills or appropriate arts and crafts. While the purpose was increasing economic output, he did not wish a return to the old system of guild masterships that kept people in an occupation to which they were ill suited. On the other hand, he said there should be no class distinction between sensory and verbal intelligence, since senses are needed for manual work, and even the highest arts like music and painting require imagination.[73]

Binet showed social prejudice in calling the son of a lawyer who became an office employee "déclassé." Nor did he refuse the prospect of classification of "normal" children despite the claims by some that his intelligence test was only intended to help the subnormal. He invited the authorities to try the "intellectual examination of young recruits before accepting them" to see "which are capable of learning the manual of arms" or "submitting to discipline."[74]

Carson points out the changes in the conception of the metric scale of intelligence as it was revised in 1908 after the initial version of 1905. The first version had no scoring or calculation of intelligence.[75] By 1908 a

69 Binet and Simon, "La Misère," 14.
70 Binet, *Ecrits psychologiques et pédagogiques*, introduction, 13.
71 Binet, *Modern Ideas*, 83.
72 Ibid., 23; *Idées modernes*, 11.
73 Binet, *Modern Ideas*, 188, 197.
74 Ibid., 94.
75 Carson, *The Measure of Merit*, 139.

new collection of tests had been reoriented from use by "subnormal" children ("idiots, imbeciles, and retarded") to teaching what "normal children" could assimilate and to designation of a normal, retarded, or advanced mental level.[76] Carson argues that in Binet's concept intelligence changed from a multivalent complex of faculties to a single faculty that could be assessed on a quantitative scale. By 1911 Binet and Simon had subdivided the mental-level score to tenths of a year and had prepared to administer the test to adults. Thus Binet himself would have paved the way for the measurement of intelligence on IQ tests. But Binet stressed that different mental faculties were involved with the various tests. "Intelligence" was a synthesis of abilities not amenable to a single table.[77] Carson astutely notes Binet's interest in craniometry and claims that Binet structured the test so that, like craniometry, it would have an instrumental and statistical approach.[78]

Nevertheless from 1905 to 1909 Binet gave little indication that he had changed his notion of intelligence, even though the synthesis was more quantified. In 1905 he had defined intelligence as "judgment, or good sense, practical sense, initiative, and the faculty of adapting one's self."[79] In his overall survey of 1909, he still defined intelligence as comprehension, adaptation, inventiveness, maintenance of focus (directedness), and ability to criticize one's own reasoning.[80] Carson argues more persuasively that resistance in the French intellectual community to quantifying intelligence and the existing selection mechanisms for elite schools by *concours* militated against the development and extensive use of the Binet-Simon test within France. Other scholars have also pointed out the resistance of the psychiatric and medical communities to an instrument considered an intrusion on their territory. No doubt Binet's critique that psychiatric tests devoted more time to elementary functions of sense and memory did not ease the adoption of his own techniques.[81]

In any case Binet himself did not believe in an absolutely fixed mental development. He claimed that "all that is thought and function in us is susceptible of development." Practice, training, and method could "augment a child's attention, memory, and judgment" so that he or she became "literally more intelligent than earlier" until an "incontestable limit" would yield diminishing returns.[82] This limit might take years to

76 Binet and Simon, "Le développement de l'intelligence," 83, 87–8.
77 Ibid., 64.
78 Carson, *Measure of Merit*, 139–47.
79 Binet and Simon, "Méthodes nouvelles," 196–7.
80 Binet, *Idées modernes*, 117–8.
81 Carson, *The Measure of Merit*, 157–9; Huteau, "La Réception," 51–65.
82 Binet, *Idées modernes*, 141–3.

reach and gains would persist. Mental exercises could therefore instill self-confidence and a concerted effort was possible to properly direct the "moral force that leads the world" rather than to concentrate on material well-being.[83] He thought that the pilot project special classes with the "abnormals" had shown real progress in improving their capacity for learning and assimilation.

Perhaps one of Binet's greatest contrasts with Ribot is his recognition that the stages of civilization thesis misleadingly attributed too much to the idea that sensory intelligence was the characteristic form of childhood and the primitives. Senses among "primitives" were developed earlier in life, and they lacked the intelligence of the civilized. But "why deprecate others" when poetry and the novel require the soul of a child?[84]

CONCLUSION

Binet's career illustrates his willingness to consider the effects of the social milieu and to give up, like Manouvrier, an unthinking devotion to physical measures. While others might argue that race mattered, Binet became convinced that poverty mattered for educational achievement. Far from being a social revolutionary, he thought schools could make allowances to improve the nutrition of the malnourished. He never despaired of improving the performance of the mentally challenged with the right stimuli, whatever the possible influence of hereditary diseases from the parents or of the lifestyles of parents, including alcoholism. But he was realistic enough to conclude there were limits beyond which educational systems could not help. A correlative conclusion about Binet could be the illustration of the great fascination anthropological assumptions always had for a leading comparative psychologist. Binet spared no effort to investigate nearly every physical, anatomical, and physiological correlation to intelligence – from head measurement to physiognomy to handwriting – even though he never studied race and did not commit himself to hereditarian interpretation. The particular anthropometric measurements valued by anthropologists formed a basis for his correlations, until he determined they were too weak to conclude very much about intelligence. The anthropological shadow was dark enough to persuade Binet to persist in this search even when anthropometry was losing its prestige among critical anthropologists. The pervasive power of anthropological assumptions for both Binet and Ribot illustrates the deeply embedded cultural commonplaces about race and

83 Ibid., 344.
84 Binet, *Modern Ideas*, 197.

evolution (for Ribot) and physical-mental correlations (for Binet) among significant leaders of the French intellectual elite. Worldwide debate would continue about the degree of genetic endowment evident in the test results obtained by Binet's successors.[85]

Beyond comparative psychology of individuals, the French intellectual climate had altered to try to institutionalize the discipline of sociology, begun by Auguste Comte, and now represented by Emile Durkheim at Bordeaux in the late 1880s and René Worms, lawyer, councillor of state, economist, and indefatigable promoter of new journals, conferences, and institutions. In non-Durkheimian sociology, where organicism had adherents, even if they did not endorse racial hierarchy, the debate would play out in parallel fashion to the wider trends observable in the *Revue philosophique.* These tendencies would show flexibility on race, relative rigidity on gender issues, and a widespread consensus approving the goals of French imperialism.

85 For a recent development in this debate, see *Psychology, Public Policy, and Law*, 11, 2 (2005).

7

The Non-Durkheimian Sociologists

THE DURKHEIMIAN AND NON-DURKHEIMIAN SOCIOLOGISTS

THE EMERGENCE OF SOCIOLOGY in specialized periodicals and institutions in the 1890s elicited a direct revolt against anthropological varieties of biological determinism and "raciology." Emile Durkheim and his collaborators on the *Année sociologique* (first published in 1898) have long had the reputation of promoting this "discovery of the social."[1] The neglected non-Durkheimian sociologists associated with René Worms suffered from the taint of "organicism" – advocating the analogy of societies to biological organisms. Later sociologists considered that this outlook was obsolete and served no purpose. The expectation of scholars is that the non-Durkheimians would be more rigid on "race" and gender. Some scholars have even associated the Worms group with individualistic social Darwinism as well as racial hierarchy.[2]

This general portrait is due for re-evaluation. The revolt of the Durkheimians against the biological now seems less definitive. We have seen how Durkheim himself recognized a certain rationality in traditional castes in previous epochs and rejected equal perfectibility of all races.[3] His early work in the *Division of Social Labour* contained biological models

1 Mucchielli, *La Découverte du social*, 144, 271–5, on Worms; Mucchielli, "Sociologie versus anthropologie raciale," 77–95.

2 On Worms, see Gissis, "Late Nineteenth Century Lamarckism and French Sociology," 69–122; on the social composition of Worms's group and the failure to advance sociological theory see Clark, "Marginality, Eclecticism, and Innovation," 12–27; Geiger, "René Worms, l'organicisme et l'organisation de la sociologie," 345–60; Mosbah-Natanson, "Internationalisme et tradition nationale," 35–62; an exception to the view that the Worms group was racist is Barberis, "In search of an object," 51–72, esp. 56.

3 Lehmann, "The Question of Caste in Modern Society," 566–85; Pedersen, "Sexual Politics in Comte and Durkheim," 229–63; Paligot, *La République raciale*, 192–3.

and organic metaphors, including citations from Gustave Le Bon on the brain size of women.[4] Durkheim's dominant gender theory argued that sexual similarity and equality in structure and function were primitive, while greater sexual differentiation occurred with modern division of labour.[5] Consequently, greater similarity of function through women working outside the home was problematic for Durkheim. Despite agreeing to legal equality for women, he thought their natural aptitudes and their functions in modern society would limit choices of occupations.[6] One might even argue that despite Durkheim's disdain for physical anthropology and biological approaches, there was a latent naturalism in his approach to gender issues.[7]

The approach to race and gender of the sociologists around the philosopher, lawyer, and political economy teacher René Worms did not consistently allow nature to trump nurture. However, they were not uniformly on one side of the nature-nurture axis with regard to race nor were their views on gender entirely congruent with their views on race. A substantial contingent among Worms's associates, including critical anthropologists, Russian émigrés, and active contributors to the *Revue internationale de sociologie*, criticized racial theory, while a minority remained committed to old anthropological concepts of hierarchy disavowed by the Durkheimians. Worms himself retained a belief in the inequality of races while allowing for substantive influences of the milieu.

While belief in racial hierarchy often helped legitimize empire, its absence did not guarantee opposition to French imperial expansion. Organicists in the Worms group viewed colonization as a necessary form of reproduction in growing societies. Other associates shared the common economic and strategic motives of French politicians and imperial theorists.

The complexity and inadequacy of an exclusive nature-nurture framework becomes even clearer on gender issues. While many non-Durkheimians did not insist on an indelible biological nature of women, their acknowledgment of socially variable gender roles did not easily translate into approval of full civil and political equality for women. In

4 Emile Durkheim, *De la Division du travail*, 24–5; Durkheim, *Le Suicide*, 57–8; Lehmann, "The Question," 570–1; Besnard, ed. *The Sociological Domain*; Lukes, *Emile Durkheim*; Blanckaert, *La nature de la société*; Hadjian, "Emile Durkheim et la taille des crânes," 19–30; Mucchielli, *La Découverte du social*; "Sociologie versus anthropologie raciale," 88–9; Mucchielli, "La dénaturalisation de l'homme," 41–53.

5 Lehmann, "The Question," 576; Charron, "Représentations des rapports sociaux de sexe."

6 DiCristina, "Durkheim's Latent Theory of Gender and Homicide," 225–6, 229, on "familial" feminism.

7 Charron, "Représentations," 4–5.

this respect, they were not so different from the Durkheimian authors of the *Année sociologique.* The Durkheimians admitted a small number of women authors (mostly German and American) among books to be reviewed.[8] The Worms group was progressive in admitting women as society members and listening to even the most radical feminists. But a strong contingent worried about childbearing in a context of depopulation if women worked outside the home and participated in the public sphere. Some still defended traditional theories of the "nature" of women. The most numerous group used social, not biological, justifications to contest radical feminist demands. The stability of households remained a major issue, as it did with Durkheim himself. In this way the sociologists illustrate with some qualifications the general claim by scholars about cultural assumptions (shared by conservatives, so not uniquely liberal or republican) creating obstacles to women's full citizenship.

René Worms established three sociological institutions – the periodical he edited, the *Revue internationale de sociologie* (founded in 1893 with Ribot as a usually inactive collaborator on the editorial board), the prestigious International Institute of Sociology (1893), with publication of annual or periodic accounts of its congresses, and the Paris Société de sociologie (1895). The International Institute of Sociology, with 100 members and 200 associates, included well-known philosophers, anthropologists, and sociologists from Europe and the Americas. Most members were university lecturers already famous in their home countries and often established in the disciplines of law, economics, history, philosophy, and anthropology. At any given annual (or later triennial) congress perhaps only twenty to thirty members and associates presented papers, although the audience, including officials, the press, and other scholars, could number 150 to 200, or even 500 in 1909 at Berne.[9] The *Revue* contributors included an assortment of Spanish and Italian scholars as well as a conspicuous circle of Russian émigrés to France who had also joined the institute. Since Britain, Germany, and the United States had their own sociological journals, they were less likely to furnish periodical authors than members of the Institute.

The Paris Société de sociologie had between 100 and 300 members in the period from 1895 to 1914. Gabriel Tarde was the first president, but the animated discussions usually occurred among a small group of ten to twenty journalists, civil servants, and the same Russian émigrés.[10] Eclipsed by the Durkheimians as founders of the discipline in France,

8 Ibid., 119–20.

9 Schuerkens, "Les Congrès de l'Institut international de sociologie," 7–24, esp. 11–18.

10 Clark, "Marginality," 20–2 and Geiger, "René Worms," 359–60.

the group, like its founder Worms, remained eclectic, without a unifying theoretical outlook in its discussions. The very success of their internationalism, without gaining legitimacy in French university institutions, seems to have surrendered the field to the Durkheimians, who were coherent theoretically and skilled in creating a French orientation to their construction of sociology.[11] Discussions of race and gender among the Worms group however, remain fascinating.

Race and gender were asymmetrical, since the advancement of European women could be more easily contemplated than the rapid development of "retarded" races. The non-Durkheimian sociologists, like the authors of the *Revue philosophique*, would end up reversing this asymmetry with more flexibility in practice on race than on gender. The most likely reason is the context of increased concerns about the legal, economic, and political status of women.

The old racial paradigm did not entirely collapse in the face of critics such as Manouvrier.[12] While the late nineteenth-century sociologists almost all attributed a lesser influence of race to the complexity of developed societies, including mixing of groups, and the phenomena of modernization, the actual difficulty of finding reliable racial indicators and demarcation lines via anthropometry was principally responsible for the critical onslaught. Neither the cephalic index nor the facial angle nor cranial size seemed to suffice as indicators. Peoples and nations were not homogeneous, and there was no firm agreement on the substratum of races.[13] Hence the fin-de-siècle era witnessed the increasing inadequacy of a nevertheless persistent paradigm.

THE RACE/GENDER PARALLEL AND DEFINING FEMINISM

As Nancy Stepan has maintained, opinions on race and gender often intersect. For many scientists "gender was found to be remarkably analogous to race, such that the scientists could use racial difference to explain gender difference, and vice versa."[14] Routinely anthropologists used alleged low brain weight or alleged emotional immaturity to draw analogies between women and men of "lower races." The exploitation of the bodies of non-European women and the dilemma of the national, racial, and cultural identity of Métis children were also issues in the common

11 Mosbah-Natanson, "Internationalisme et tradition nationale," 42–7.

12 Blanckaert, "La Crise de l'anthropométrie," 95–172; Hecht, "A Vigilant Anthropology," 221–40; Hecht, *The End of the Soul*, 28, 211–56.

13 Paligot, *La République raciale*, 79, 83, 281.

14 Stepan, "Race and Gender," 261–77, esp. 263.

denigration of certain "races" and the feminine.[15] Yet the male sociologists did not always parallel race and gender. In some cases, conservatives and liberals were consistent on both issues, while the views of others attacked social conventions on one issue but not the other.

Any discussion of feminism in this era must also acknowledge the problem of definition. The most popular view among both men and women members of the Société de sociologie de Paris was the "relational feminism" retrospectively defined by Karen Offen. Rather than being focused on a gender-free abstract individual bearing rights, relational feminism challenged male domination by the "primacy of a companionate, non-hierarchical, male-female couple" with significant support for childbearing, nurturing, and "womanliness."[16] This view allowed for a "biologically differentiated, family-centred vision of male-female complementarity," a "sexual division of labour in both society and the family."[17] Joan Scott has articulated the long-standing paradox within the feminist movement – the more women emphasized their difference with special concerns about motherhood and household, the more difficult it became to argue for the extension of universal rights.[18] However, Scott counsels against a forced choice between equality and difference as a policy. Offen has argued against excluding more conservative feminists who chose complementarity.[19] Clearly both individualist and relational feminists participated in the sociologists' discussions.

The difficulty of conceiving non-Europeans as well as French women as republican citizens has been a staple of recent historical scholarship.[20] The scandal has been the paradox of liberal ideals and illiberal behaviour. However, a recent study of family legislation has complicated the picture by showing divisions among republicans themselves on the degree of responsiveness to feminist demands.[21] The male adherents to the Société de sociologie de Paris did not have foreordained opinions about gender. They demonstrate a range of opinions from support of full civil and political rights for women to limiting women to the domestic sphere. Most seem to have embraced cultural assumptions about the household

15 Stoler, "Sexual Affronts and Racial Frontiers," 198–237.

16 Offen, *European Feminisms, 1700–1950*, 22–3.

17 Offen, "Depopulation, Nationalism, and Feminism," 648–76, esp. 664–7.

18 Scott, *Only Paradoxes to Offer*, 1–18, esp. 3–4.

19 Offen, "Defining Feminism," 119–57; comment by Cott, 203–5, reply by Offen, 206–9.

20 For the argument that republicanism inevitably excluded women see Sowerwine, "The Sexual Contract(s) of the Third Republic," 245–53; for non-Europeans see Conklin, *A Mission to Civilize*.

21 Pedersen, *Legislating the French Family*.

common not to just to adherents of liberal republican ideology but to conservatives and anti-republicans.

RENÉ WORMS ON RACIAL HIERARCHY AND COLONIZATION

In a way, Worms represents the split personality of his group. His conflicted view of race considered it an important, though not predominant, variable. In 1895 he criticized the conservative, pessimistic, anti-republican racial theorist Arthur de Gobineau's work as an "effort to glorify the white race."[22] In the same year he commended the republican philosopher Alfred Fouillée for opposing "doctrines too widely spread today about the radical inferiority of certain fractions of the human species and the progressive elimination of more enlightened races by the inferior fractions." Three years later he also agreed with Fouillée, in a very commonly expressed outlook akin to that of Gaston Richard among the early Durkheimians, that "the more a people approaches a modern type, the action of the social milieu wins out over the physical milieu, and even more, physical factors tend to be transformed into social factors."[23] In a 1902 discussion at the International Institute, Worms warned his colleagues not to attribute too much to race: "Milieu is a factor of social evolution, modified by social functioning, while race cannot be modified and cannot modify other factors. It can as a conservative factor only prevent transformations."[24]

For someone eager to highlight the influence of milieu rather than heredity, Worms stubbornly refused in later years to give up the racial paradigm. Somatic structure was a measure of progressive adaptation to the environment. "To the extent one rises in the series of human races," he argued, cranial volume and brain weight increase, the facial angle increases, and prognathic (receding forehead and prominent jaw) face conformation decreases. To some degree Worms shared the preconceptions of Ribot about savages and was more dogmatic about race than the Durkheimians. In trying to "modify the mentality of Negroes by a European education" Worms believed there were "countless disappointments," such as the return of a dark-skinned Filipino to savage life after failed efforts to civilize him. He attributed the failure of democracy in Liberia and Haiti to racial inferiority. On the other hand, he conceded

22 Worms, *Organisme et société*, 34–5, 159, 307–8.

23 Worms, Review of Fouillée, "Tempérament et caractère," 1059; Review of Fouillée, "Psychologie du peuple français," 658–9.

24 Worms, Discussion of La Grasserie, *RIS* 10 (1902): 384–5.

that the remarkable recent progress of the Chinese and Japanese might herald a "transformation more profound than we can now believe."[25]

Worms's nuanced view on the importance of race did not prevent him from encouraging the "praiseworthy task of development of our far-flung possessions by sending workers and capital" to the colonies or from acting as secretary of the Colonial Congress in 1905.[26] Nor did the Dreyfus affair make Worms more receptive to the arguments of the noted Durkheimian critic of racial theory Célestin Bouglé.[27] Somewhat like Espinas and Jankélévitch among the authors of the *Revue philosophique*, Worms warned that Bouglé's exclusion of physiological and psychological methods from sociology "would restrain the resources of the science."[28]

Worms also permitted the publication of the views of the anti-democratic extremist "anthroposociologist" Vacher de Lapouge.[29] In 1893 Lapouge argued, contrary to Broca, that the effects of education were limited and did not enlarge the brain.[30] He advocated eugenic measures to ensure that war, urban-induced alcoholism and disease, marriages to the merely wealthy, or healing the mediocre sick would not threaten the dominance of the elite. The *Revue* published discussions of the cephalic index by his disciples. In the first years of the *Année sociologique*, of course, even the Durkheimians allocated space to Lapouge, though they removed the "anthroposociology" heading by 1901.[31] One cannot however, conclude that Worms's institutions were mouthpieces for extremist advocates of racial hierarchy. A more subtle interpretation emerges from three substantive discussions of race and heredity.

25 Worms, *Principes biologiques de l'évolution sociale*, 50–1.

26 For the classic, still helpful explanation of imperialist ideology, see Betts, *Assimilation and Association*, 88, on Worms; for the ethnographic study of colonial populations, see Chauffard, "Les Populations indigènes du Cambodge et du Laos," 561–6; Bernard, "Les populations de la vallée du Mekong," 567–89; Caillaud, "Les populations du Cambodge," 590–6; citation from Worms, Review of "La Quinzaine coloniale," 238.

27 On Bouglé, see Hecht, *End of the Soul*, 264–9; Vogt, "Un Durkheimien ambivalent," 123–39; Logue, "Sociologie et politique," 141–61; Logue, *From Philosophy to Sociology*, 180–94; Policar, "Science et démocratie," 86–101.

28 Worms, Review of Bouglé, "Les idées égalitaires," 315–16; Worms, Review of Bouglé, "La Démocratie devant la science," 817.

29 Taguieff, *La Couleur et le sang*,199–325; Hecht, *End of the Soul*, 168–210; Massin, "L'anthropologie raciale," 269–334.

30 Vacher de Lapouge, "Le Darwinisme dans la science sociale," 175–85 and 414–36; see also the American Carlos Closson, "Dissociation par déplacement," 211–22, the French disciple Henri Muffang, "Etudes d'anthropo-sociologie," 789–803, and the German Otto Ammon, "Histoire d'une idée," 145–71.

31 Mucchiellli, "Sociologie versus anthropologie raciale," 88–9.

RACIAL HIERARCHY IN THE SOCIOLOGY SOCIETY AND INTERNATIONAL INSTITUTE

At the 1895 second annual Congress of the International Institute the Polish expatriate, anarchist pamphleteer, and literary critic Mécislas Golberg (1868–1907) highlighted the importance of milieu in more recent times by asserting that the "stable morphological units" of races in early times adapted to their geographic milieu to create a division of labour. Conquest and enslavement of one race by another was one possible outcome, but another, as Durkheim suggested, was greater specialization leading to natural commerce and co-operation. In historical times Golberg seemed to conflate races and nations by asserting that races are "subjective and unstable social units" since nations are unions of different peoples.[32]

A recalcitrant hard-line critic of Golberg was a former Proudhonist typographer, Charles-Mathieu Limousin (1840–1909), a councillor and eventually officer of the Société de sociologie de Paris. For Limousin, racial attributes were permanent and the hereditary constitution of Africans made progress toward civilization impossible.[33] In the subsequent discussions of 1900 and 1902 Limousin again insisted that the once useful ideals of equality had no place in a science of society where one could not contest the "differences of social aptitude for each race." Asians were in a state of arrested development, and Native Americans in a "state of decadence."[34] Limousin represented almost the stereotype of the "reactionary Left," a type of intellectual steeped in French Revolutionary values harking back to Jacques-René Hébert, with an opinion receptive to nineteenth-century socialism and a firm conviction that progressive science taught hierarchy, not egalitarianism.[35]

However, at the 1895 congress, after the conviction of Dreyfus but before the press helped create the Dreyfus affair of 1898–99, three major figures, all active in the Société de sociologie de Paris, strongly supported the primacy of social, not racial, factors. The cordage manufacturer

32 Golberg, "L'Origine des races et la division du travail," 345–85, esp. 348, 350, 352, 355, 358, 360.

33 Limousin, Discussion of Golberg, *AIIS 1895* 2 (1896): 368; on this exchange see Mucchielli, "Sociologie versus anthropologie raciale," 89, and on Worms's Institute, *La Découverte du social*, 145–50.

34 Limousin, "Le Progrès inégal des races," 261, 267; Limousin, Discussion of Roberty, "Les Préjugés de la sociologie contemporaine," 259–67; see also Limousin, Discussion of La Grasserie, "La Race et le milieu," 392–3.

35 See Crapez, *La Gauche réactionnaire*, for the careers of Albert Regnard and Jules Soury, and regarding Boulangism as a distant forerunner of fascism; Taguieff, *La Force du préjugé*.

from Odessa Jacques Novicow (1849–1912) was a staunch organicist, social Darwinist, and economic liberal, but also a strong pacifist. For him, Golberg's adoption of Le Bon's "historical races" confused race and culture. The physiological aspects of race, however useful in biology, were unimportant in sociology. Therefore, sociologists should "abandon this criterion of race for tracing serious limits between human groups." He remarked on his own sensitivity to prejudice because others frequently assumed him to be Jewish, but he had no such ancestry.[36] The next year Novicow argued against any correlation of intelligence with brain size or physiology, and found no evidence that primitive races or present-day Africans were not "perfectible."[37] Novicow would also turn out to be an ardent supporter of equality for women.

The statistician, economist, and future president of the society, Adolphe Coste (1842–1901), asserted that marriages, uniform traditions, and education were far more important for group mentality than physical attributes. Physiology had no direct influence on mental characteristics, and, like most of his colleagues, he believed all existing races were mixed.[38] Finally, in 1895, four years before his attack on Lapouge, and nine years after his lukewarm reaction to the Haitian Anténor Firmin's plea for equality, Manouvrier stressed the importance of "conditions of civilization" attainable by all races. Rejecting "original and inherent features of a race," Manouvrier wished to react against "the habit of explaining by race, blood, heredity, and atavism what is explicable by the external milieu and the action of living beings assembled together."[39] He had little doubt, he said, that on the average "as a whole the Europeans are superior to exotic races" but he thought culture, education, and favourable conditions could counteract current inferiority. Manouvrier later admitted the cerebral inferiority of some existing human "races," but still rejected Le Bon's serial ranking of intelligence testifying to progressive evolution.[40] We have seen how his position on feminism was similarly nuanced – he did not believe in the inferior intelligence of women, but was convinced that women's nature meant adherence to certain social roles.

36 Novicow, Discussion of Golberg, 368; cf. Novicow, Discussion of Roberty, 272.

37 Novicow, *L'Avenir de la race blanche* (1897), cited in Paligot, *La République raciale*, 145–6.

38 Coste, Discussion of Golberg, 369; Coste, Discussion of Roberty, 271–2.

39 Manouvrier, Discussion of Golberg, 379–80, 383, 385; cf. Paligot, *La République raciale*, 54–6.

40 Manouvrier, "Le progrès anthropologique de l'intelligence," 77–100.

The second explicit discussion on race in sociology occurred at the 1900 International Institute Congress well after the Dreyfus affair had galvanized concern about prejudicial labelling of individuals. Here another nobleman and wealthy émigré, Eugène de Roberty (1843–1915), who still lived half the year in Russia and also taught at the Paris Collège libre des sciences sociales before his return to St Petersburg in 1904, vigorously attacked the "prejudices of contemporary sociology."[41] The positivist Roberty blamed politically conservative psychologist and historian Hippolyte Taine, British philosopher Herbert Spencer, the two stars of the early *Revue philosophique*, as well as republican moderate or environmentalist philosophers Jean-Marie Guyau and Fouillée, for overemphasizing organic adaptations transmitted by heredity. Despite the significance they attached to race, it was actually a deus ex machina with the "external veneer of science." Even so-called "ancestral concepts" were acquired. As long as education and instruction were different from raising animals, he believed race itself was not a great social factor. Furthermore, he called anti-Semitism "the greatest infamy the nineteenth century will have to blush about before posterity."[42] Race was a biological fact, good only for prehistory, but otherwise civilization was a more essential sociological factor than race.

In the subsequent discussion, Novicow and Coste endorsed Roberty's opinions. Coste insisted Egyptians were a black African civilization, and deliberately cited the failure of Lapouge's cephalic index as a successful indicator of racial difference.[43] The Breton magistrate, sociologist of law, and South American linguistics expert Raoul de La Grasserie (1839–1916) contradicted Roberty in maintaining that nations were now artificial races – a principal premise of his view that all true nations deserved self-determination. The analyst of Russian peasantry Maxime Kovalewsky supported the conventional anthropological assumption that all races evolved along the same path, but that some were in a state of arrested development. Limousin adhered to his hard-line position about the pure sentimentality of egalitarianism.[44]

41 Roberty, "Les Préjugés de la sociologie contemporaine," 239–59; discussion, 259–72; on Roberty, see Yusef Semlali, "Eugène de Roberty (1843–1915)," http://halshs.archives-ouvertes.fr/halshs-00003964/en/; for his corresponding membership in Littré's earlier Société de sociologie, see Heilbron, "Sociology and positivism in 19th-century France," 38, 53.

42 Roberty, "Les Préjugés," 245, 249–50, 252.

43 Coste, Discussion of Roberty, 271–2.

44 Limousin, Discussion of Roberty, 259–60; La Grasserie, Discussion of Roberty, 267; Kovalewsky, Discussion of Roberty, 268–70.

In the third discussion of this theme, at the Société de sociologie de Paris in April 1902, La Grasserie translated race as an "ethnic" quality anterior to society, necessary for "anthroposociological practice," a term used by Vacher de Lapouge.[45] He agreed this ethnic quality was most powerful in earlier eras and unmixed groups. However, he refused to surrender a firm naturalistic component – poor ethnic stock would hamper a people regardless of the two other dimensions of geographic milieu and historical epoch. La Grasserie postulated a perpetual struggle between a hereditary ethnic factor (a "better name" for race, given their mixing), remade by history and the geography of soil and climate. At present the nations, or "sociological races," were more significant, but nations now took on the characteristics of races. Like Le Bon, he unabashedly argued that in his view the "ethnic" characteristics of Jews, including an aptitude for international commerce and finance, always triumphed over the "telluric" (geographic) milieu.[46]

In the discussion Worms took his usual conciliatory position by cautioning that while race was an immutable factor important in the past, only the milieu promoted social evolution. Limousin again endorsed La Grasserie's argument for the importance of race. The anthropologist Georges Papillault, a laboratory colleague of Manouvrier, insisted on the old paradigm – a parallel between racial morphology and functional aptitudes in an evolutionary series. In a given epoch different races reacted differently to the same excitation, so the influence of the milieu was limited, though important. However, he tried to separate his science and his ethics – there was no excuse for savage destruction of an inferior race.[47]

The increasing disunity at the Ecole d'anthropologie appeared in the critique of the cartographer Franz Schrader, who was a colleague of Manouvrier. Schrader argued that the constantly changing nature of "races" robbed the word of its meaning. The famous criminologist and magistrate Gabriel Tarde pursued the same line of reasoning as in the *Revue philosophique.* He agreed, like Durkheim and Worms, that as evolution continued, the ethnic factor decreased in importance, while "continually operating causes" such as social life increased. He concluded peremptorily, "The idea of exclusive or preponderant influence of race leads, at bottom, only to a historical fatalism … a sociological mysticism." Moreover, the genius, not the average person, promoted social

45 La Grasserie, "La Race et le milieu," 356–94.

46 Ibid., 357, 359–60, 365, 378, 382–3; see also La Grasserie, "Principes sociologiques des nationalités," 426.

47 Papillault, Discussion of La Grasserie, "La Race et le milieu," 387–90.

evolution, so that differences in racial averages had relatively little significance for the establishment of civilization.[48]

Hence Limousin and Papillault, and to a lesser degree Kovalewsky, defended the traditional anthropological paradigm relating physical natural attributes, intellectual aptitudes, and moral character. La Grasserie postulated a pre-existing ethnic factor that could strongly affect adjustments to the milieu. Worms remained in an intermediate position, refusing to deny or magnify the influence of race. Manouvrier and Schrader from the Ecole d'anthropologie, along with Novicow, Coste, and Tarde declined to attach great significance to racial variables alone. For Coste and Manouvrier, at least, the failure of anthropological indicators was a major influence.

Was there another argument that attracted all the participants, as well as Durkheimians: the increasing importance of social factors as ethnic groups became more mixed? It would be a tempting reading of these debates to say that processes of modernization in Europe in the late nineteenth century made it impossible to defend a fixed racial category. But modernization did not eliminate the old paradigm. Some anthropologists still questioned whether Métis of allegedly "distant" races would result in the deterioration of the species.[49] Moreover, we shall see how an ideology of indelible races, albeit leavened with cultural factors, remained alive and well in the 1930s and into the Vichy era. Hence it is still speculative to attribute the fall of the old paradigm to an overarching process of modernization.

THE *REVUE INTERNATIONALE DE SOCIOLOGIE* ON RACE AND ETHNIC IDENTITY

Aside from these explicit discussions, the book reviewers of Worms's periodical demonstrated a mostly critical attitude toward racial theory. However, society discussions or publications about African-Americans and Jews were less clear-cut. The workhorse reviewers after 1900 included the lycée philosophy teacher Guillaume-Léonce Duprat and the social democratic lawyer Alfred Lambert. The most far-reaching disavowal of conventional anthropology was Duprat's review of the Austrian sociologist Ludwig Gumplowicz's theory of conflict among historical races. While Duprat believed in hereditary dispositions, he noted, "we find contestable the hypothesis of the natural existence of race."[50]

48 Ibid., 385–7.

49 Blanckaert, "Of Monstrous Métis?" 42–69; Zerilli, "Il Dibattito sul meticciato," 237–73.

50 Duprat, Review of Gumplowicz, "La Lutte des races," 542–4; cf. Duprat, "La Psychosociologie," 656–77.

In striking contrast to the philosopher Henri Joly in the *Revue philosophique*, Lambert favourably reviewed the noted work of the naturalized Polish journalist Jean Finot, who attacked the racial theories of Gobineau and Vacher de Lapouge. Lambert approved Finot's strictures against the "harmful and vain doctrine of races," an "intransigent dogma unleashed by political passions" and not scientifically justified. He concluded, "the cultivation of hatred of races is criminal; internally it would be a seed of death for peoples that would let it take root; externally, it would lead to the worst sort of decline; it is therefore a human and patriotic desire fulfilled by Jean Finot in chasing from the scientific throne a disastrous usurper."[51]

However, the direct comments on African-Americans and Jews in the *Revue* were far from uniformly sympathetic. La Grasserie failed to criticize an American statistician's pessimism on educating African-Americans.[52] A former resident of the American South, Mrs Oscar Lovell Triggs, one of the few women who participated in the debate on race, told the society that blacks would never achieve the superior qualities of whites.[53] Still the historian, economist, and commentator on Haiti, Paul Vibert (1851–1918), married to a West Indian woman, argued that the "savage mentality" of poor white Americans handicapped blacks. In the same discussion, the Haitian embassy official and lawyer Georges Sylvain deplored the Americans' maintenance of segregated hotels even in New York.[54]

The feelings aroused by the Dreyfus affair produced several comments on the status of the Jews. Duprat's review mentioned some of the replies to the writer Henri Dagan's inquiry into the causes of anti-Semitism. Manouvrier, in particular, saw no inherent "pretended flaws" in the "pretended" Jewish race.[55] Clearly Worms, himself of Jewish ancestry, at one time rejected the notion that the Jews were a race, but he still later referred to the psychic transformations of Jews as representing one group "among white races."[56] Neither the Worms group nor the Durkheimians wished to be associated with Le Bon's nationalist anti-Semitic views or Vacher de Lapouge's Aryanism.

51 Duprat, Review of Finot, "Le Préjugé des races," 907–8; on Finot, see Hecht, *End of Soul*, 270–2.

52 La Grasserie, Review of Hoffman, "Des Traits et des tendances," 137.

53 Triggs, "La Question nègre," 197; see also Gagey-Macadoo, Review of Dubois, "Efforts for Social Betterment among Negro-Americans," 699–701.

54 Vibert and Sylvain, Discussion of Louis Skarzynski, "Les Peaux-Rouges et les Nègres aux Etats-Unis," 355, 362.

55 Duprat, Review of Dagan, "Enquête sur l'antisémitisme," 909–10; for Manouvrier, ibid., 910.

56 Worms, "Le Concept de la société," 81–93, esp. 88n; *Principes biologiques*, 52–3.

Anti-Asian prejudice dominated the commentary of labour law expert Léon Douarche.[57] However, Novicow was strongly opposed to any notion of a "yellow peril," because he was confident that Chinese labour competition could not easily undermine European productivity.[58] On the whole the editorial staff of the *Revue*, including Lambert and Duprat, criticized the extremist anthroposociology of Lapouge and discarded conventional anthropological assumptions. But the old prejudices survived in the more casual contributors to sociological discussions on Asians or Africans.

HEREDITY AND PROFESSIONS

While the subject of heredity and the professions is not directly related to either race or gender, it forms another chapter in the relative role of nature and nurture that affected all of these debates. Francis Galton and Ribot had insisted on a hereditary component in choice of professions, while even Durkheim had conceded that traditional caste societies had a natural substratum. In 1910 Worms remained convinced that inheritance of intellectual aptitudes in family dynastics was important, though he could not determine the relative influence of education. Moreover, after a series of generations, acquired characteristics could be embedded in the organism. However, like Durkheim, he thought the influence of heredity declined in modern society.[59] Worms's own sampling for the legal profession in 1899 showed that about half the Paris bar had fathers connected to law, while all families sought to ascend to more prestigious positions.[60]

Coste had previously argued that large families would provide more opportunities for achieving a desirable transmission of professions to the next generation (unlike Durkheim, who thought such transmissions hindered progress).[61] He now found that out of ninety-seven Frenchmen, twenty-two had the same profession as their father, twenty-five a closely related or similar profession, while fifty were in an unrelated field. But both Coste and the socialist demographer Arsène Dumont agreed that imitation, education, and the stability and prosperity of the field all influenced professional choice. The physician Fernand Gaucher suggested that the real subject should be heredity of aptitudes, not professions, especially when mechanization was rapidly displacing artisans.[62]

57 Douarche, "La Vie sociale aux Etats-Unis," 193–4.
58 Novicow, "Le Péril jaune," 351–5.
59 Worms, *Principes biologiques*, 70–4.
60 Worms, Société de sociologie, "Les Bases sociales," 53, 122.
61 Coste, *Les Conditions du bonheur*, 32–7; Durkheim, *De la division du travail*, 291–2.
62 Société de sociologie, "L'Hérédité des professions," 50–1, 53, 55, 59, 117.

Limousin, who was a strong advocate of racial theory, argued that psychic heredity for professional aptitudes was not the same as physical heredity. In view of changing social conditions, the concept of heredity of professions was an error of "secondhand scholars" who tried to transform a physical law of heredity into a psychic law. Similarly, Léon Philippe, a bureaucrat in the Agriculture Ministry and a political economy and statistics teacher, joined the general consensus that general aptitudes, not specific professional skills, could be inherited.[63] Most society members realized that social values such as owning land or technological changes could influence the careers of children. Hence, the debate on hereditary professions helped to unseat comfortable clichés about transmission of specific callings in favour of general dispositions amenable to social influences.

COLONIZATION

Critiques of racial theory rarely expressed anti-imperialist sentiments. Conservatives such as Le Bon and Jules Harmand were as vocal as liberal republicans in using racial hierarchy to justify empire.[64] In this era of expansion in equatorial and west Africa and consolidation of control in Indochina, a paternalist ideology assuming the need of "less evolved" peoples for guidance predominated. The sociologists did not employ the universalist and assimilationist arguments cited by Alice Conklin as sometimes convincing to paternalist republicans – that is, bringing law and order and combatting aristocracy and slavery among these peoples.[65] For organicists such as Worms, colonization was an inevitable manifestation of the need of highly developed societies to reproduce.[66] His participation in the Colonial Congresses testified to his keen interest. The strongest pro-hereditarian statement applicable to colonization came from the physician and librarian to the Navy Ministry Charles Valentino. He warned against destruction of caste systems in the empire as a flaw in "social economy." Education could only deform heredity, not seriously change it.[67] Hence, in line with the "associationist" (as contrasted to "assimilationist") imperial ideology that gained ground around 1900, he would advise colonizers to preserve the institutions of indigenous peoples, though they might be repugnant to the values of the imperial power.

63 Ibid., 117, 120, 121–4, 206–7; see also contributions of Prince Ténichef, Louis Favre, Henri Rappin, Ernest Delbet, and Louis-Léger Vauthier.

64 Paligot, *La République raciale*, 241, 253.

65 Conklin, "Colonialism and Human Rights," 419–42.

66 Worms, *Principes biologiques*, 75.

67 Valentino, "L'Evolution de l'intelligence sous le régime des castes," 179, 191.

However, Vibert earnestly questioned the practices, if not the premises, of colonization. In 1910 after Mme Jeanne de Maguerie (another woman who participated in debates not about feminism in itself) recommended including women in the civilizing mission to uncover the secrets of the harem in north Africa, Vibert warned that colonizers would reap the hatred of their subjects if they treated supposedly inferior peoples with harshness, brutality, and injustice. "There are no inferior races on earth, only more or less advanced." Yet Vibert thought that even "advanced" peoples such as Annamites (Vietnamese) and Arabs needed protection and benevolence. Even as an opponent of racial hierarchy, he never flagged in his support for the economic motives of colonization. Indeed, he believed that the future salvation of Europe depended on development and "rational exploitation of colonies."[68]

Even socially radical members of the society seemed well disposed toward colonization. The young Marxist noble from Poland, Baron Casimir de Kellès-Krauz, insisted that a practical application of sociology could be the justification of colonization for worldwide progress and civilization.[69] The social democratic lawyer Lambert, also a critic of racial theory, reviewed in 1899 the most famous exposition of "associationist" imperial policy by the retired naval officer Léopold de Saussure, *Psychologie de la colonisation française*. Lambert noted without criticism the need to understand "psychological races," and to replace failed assimilationist ideals with a better colonial policy that would avoid revolt. Thus, security and road building must precede the teaching of French and reform of the educational and administrative systems. According to Saussure, one must, like the British, respect indigenous customs as long as they do not violate public morality. Lambert never questioned the delay and perhaps permanent refusal to grant indigenous peoples the rights of republican Frenchmen. Some years later, Lambert's review of the journal *L'Année coloniale* commented that modern colonial policy displayed comparative sympathy for indigenous peoples, despite isolated revolts.[70]

Similar support for "associationism" came from the legal historian and appeals court lawyer Marcel Pournin, who reviewed the fifth edition of the economist Paul Leroy-Beaulieu's *De la Colonisation chez les peuples modernes*. He, too, pointed out the need to respect native practices, so that even radical abolition of slavery might be postponed to allow

68 Société de sociologie, Discussion of "Le Colonial, type sociologique," 132; cf. Vibert, *La Colonisation pratique et comparée*, 1: 9–10, 27, 254.

69 Kellès-Krauz, Discussion of Worms, "La Distinction entre l'art et la science," 179–80.

70 Lambert, Review of Saussure, "Psychologie de la colonisation française," 553–4; and Review of L. Brunel, Ch. Mourey, and Chemin-Dupontès, "L'année coloniale," 169.

for gradual, progressive evolution.[71] While a few members openly declared their refusal to extend citizenship to the colonized, no one questioned French domination in the allegedly civilizing mission. The associationists retained racial hierarchy to legitimize imperial rule, and opponents of racial theory among the sociologists invoked economic motives or the need for strategic security to overcome the paradox of a republic with imperial subjects.

DISCUSSION ON GENDER ROLES

The Société de sociologie de Paris debated the social rôle of women at its meetings and in the *Revue*, with the participation of feminists. Using Karen Offen's definition of feminists, both moderate "relational" feminists and radical "individualist" feminists were among them. Since the 1880s women's organizations had campaigned for the opening of schools and careers to women and for protection for women workers and young women sold into prostitution. Objectives for reform of the Civil Code at first included legalization of divorce and then the facilitation of divorce by mutual consent or in some cases at the demand of one partner. Feminists also wanted legal recourse against abusive husbands and the right to launch paternity suits. Ultimately, especially during the first decade of the twentieth century, more and more women campaigned for the right to vote. Only a few of the most radical feminists raised issues of reproductive freedom.[72] This period witnessed several major congresses dealing with women's rights. The glass could be perceived as half full or half empty. In the words of one historical survey, "equality was on the march," yet progress toward each objective was glacial and political rights did not materialize.[73]

On these issues the divisions among male sociologists reflected those in society at large. They held no predetermined belief in confinement of women to home and family but they also did not usually endorse unrestricted participation in the public sphere. Only the positivists and several anti-feminists voiced rigid notions of women's "essential nature," which precluded professional opportunities or civil rights. The concern for depopulation and the childbearing nature of women remained a subtext for debates on the social status of women. The most popular view expressed sympathy for the "relational" kind of feminism (or what Judith

71 Pournin, Review of Leroy-Beaulieu, "De la Colonisation chez les peuples modernes," 373–4.

72 For an overview, see McMillan, *France and Women, 1789–1914*; Hause, "More Minerva than Mars," 97–113; Pedersen, *Legislating the French Family.*

73 Klejman and Rochefort, eds, *L'égalité en marche.*

Surkis calls "conjugal complementarity") that ensured social stability, even when they favoured law reform or suffrage for women.[74]

Jacques Novicow, a former vice-president of the International Institute, was an example of a critic of racial theory who was also an ardent feminist. He lamented the lack of freedom of movement and educational opportunities for women, the inequities in disposition of property, fortunes, and legal status, and in salaries for the same work.[75] He also thought that professions and offices should be open to women.[76] In a specific attack against the anthropologists' contentions, Novicow thought it arbitrary to say brain weight was significant for intelligence. Citing Ludwig Büchner, he argued as Manouvrier had earlier, that some women's brains were larger than male brains and their relative weight proportionate to body mass was actually greater.[77] He foresaw no problems with political rights for women, and predicted that more science, art, and literature would be produced with the development of women's talents and that civilization would be enriched.[78] In his view social science, physiology, and psychology all led to this conclusion of egalitarian treatment of women.[79]

The sensational part of Novicow's contentions, however, was the honouring of all sexual unions without marriage or other social regulation. He thought such freedom would eliminate seduced and abandoned women and eliminate the problem of inaccessibility of divorce due to the expense.[80]

Lambert's review, however, showed that not all critics of racial hierarchy were equally progressive on gender. Lambert warned that the "sexual nature of women does not permit full intellectual development" or aptitude for all occupations. Moreover, some "division of tasks would always be necessary both within and outside the family" with motherhood being the most noble task. He was also scandalized by the idea of discarding the marriage contract.[81]

We have seen in a previous chapter how, by 1909, Manouvrier, had become wary of granting full equality to women, possibly because of the visibility of the feminist campaign. Maternal and familial roles of women

74 Surkis, *Sexing the Citizen*, 3.
75 Novicow, *L'Affranchissement de la femme*, 23, 25–6, 221.
76 Ibid., 29–30, 224.
77 Ibid., 48–9.
78 Ibid., 249–51.
79 Ibid., 265.
80 Ibid., 218.
81 Lambert, Review of Novicow, "L'Affranchissement de la femme," 339–41.

would be complementary to the male provider, whereas rivalry with men in all professions would result in childlessness or neglected households.[82]

The anthropologist Charles Letourneau, another member of the International Institute of Sociology, had asymmetric views on race and gender, but more flexibility on gender. In writing on the future of women just before his death, Letourneau indicated that he saw no inherent reason to assume the psychological or physiological inferiority of women. Females were not essentially different from males in animal and primitive societies, so that the characteristic "feminine mentality" was a social fact due to the division of labour and differences in occupations.[83] By Lamarckian inheritance of acquired characteristics, these changes resulted in anatomical, physiological, and psychic differences. Women were less likely to hunt or go to war, as they no longer practised these activities. But the subordination of women, as argued by John Stuart Mill, was now an obstacle to the progress of humanity.[84]

Letourneau conceded in time-honoured fashion that women were more emotional than men now, and inclined to impulsive action and intuitive judgments.[85] But he foresaw a bright future of progress in equality in education, marriage and the family, and society at large. Women would develop all their aptitudes without constraint. Property, labour, and political organization would be transformed so that the "social, mental, and physical differences between the sexes will be attenuated."[86] Lambert's review, however, reiterated his assessment of the necessity of dividing household tasks.[87]

Lambert also reviewed without criticism the companion volume to Jean Finot's essay on racism, *Le Préjugé et le problème des sexes*.[88] Finot had argued that female physiology gave no indication of a less robust intellect, less creative genius, or moral aberrations. Women's suffrage did not promote divorce in his view. The "new woman," much criticized in France, would develop her intellect and conserve her beauty in a renewal of the ideal of the eternally feminine, while she would still fulfill the roles of mother and spouse. However, Lambert failed to comment critically on the rather more restrictive view of the positivist prison director Paul Grimanelli, who had argued that women should remain primarily devoted to the family, and that working in industry should be recognized as unhealthful and

82 Manouvrier, "Conclusions générales sur l'anthropologie des sexes," 44–51, 54, 57.
83 Letourneau, *La Condition de la Femme*, 501–2.
84 Ibid., 499.
85 Ibid., 502–3.
86 Ibid., 507–8.
87 Lambert, Review of Letourneau, "La Condition de la femme," 327–30.
88 Lambert, Review of Finot, "Le Préjugé et le problème des sexes," 738.

undignified for women.[89] Grimanelli recognized that single women would need an occupation, but he clearly rejected political functions for a woman as "altering her characteristic qualities and corrupting the best and surest source of her influence. " Auguste Comte's ideas of women as moralizing arbiters of men's projects clearly affected Grimanelli.[90]

La Grasserie was not especially liberal either on race or gender issues. He opposed legal discrimination against women, but kept the traditional stereotype of the moralizing woman. Compared to men, women would be thriftier, more honest, and less alcoholic and might calm class struggles and help prevent warfare.[91] La Grasserie found divorce harmful, and thought women should remain married and modest.

The first woman's intervention came with the report of the moderate feminist Clotilde Dissard, wife of a physician member of the society and editor of *La Revue Féministe*, on the sessions of the 1896 International Feminist Congress.[92] The next year, her prediction that the future would bring equal civil and political rights for women provoked Tarde to insist without elaboration that there must be sexual differentiation in social functions.[93]

The Société de sociologie meeting of November 1905 touched off a lengthy debate on the social role of women. During the previous year, the centennial celebration of the Civil Code had roused a massive feminist counter-demonstration highlighting the injustices in the Code. A disciple of the very conservative sociologist of families Frédéric Le Play and avid pro-natalist, the civil engineer Emile Cheysson (1836–1910), a political and social economy professor at the Ecole libre des sciences politiques, proclaimed that women are equal but distinct beings. As wives of employers or proprietors, they should show appropriate kindness and tact to employees and tenants. Women of all ranks should learn domestic hygiene and the promotion of good order in the household. Jeanne de Maguerie, a correspondent of the suffragist Cécile Brunschvicg, responded that learning only household tasks was commendable as a first step, but insufficient and infantilizing for women. In any case wage-earning women needed higher wages and better education.[94]

89 Lambert, Review of Grimanelli, "La Femme et le positivisme," 391–2; see also Grimanelli, *La Femme et le positivisme.*

90 Grimanelli, *La Femme*, 104, 106–7.

91 La Grasserie, "Le Rôle de la femme," 127–8, 132.

92 Dissard, "Le Congrès féministe de Paris en 1896," 537; Klejman and Rochefort, *L'Egalité*, 105, describes her as a believer in "complementarity" and an opponent of "egalitarian" feminism.

93 Dissard, "L'Avenir du féminisme," 937.

94 Cheysson, "Le Rôle social de la femme," and Discussion, 882–3, 885; on his pronatalism, see Camiscioli, *Reproducing the French Race*, 33.

Madeleine Pelletier (1874–1939) added that merely running a household and assisting husbands' careers would maintain women in an inferior condition. Women could thus neither develop their reason nor demonstrate interest in social and political issues.[95] Pelletier's participation is remarkable because she was a physician, former anthropological researcher, and later psychiatric intern. She personified "integral" individualist feminism rather than conformist gender roles, and at the time was a socialist in the Marxist faction of Jules Guesde.[96]

The historian Hippolyte Monin (1854–1915), also a conservative on racial issues, supported Cheysson by pointing out the danger of abandonment of the home and family, the "element of society." Women could not have the same rights and duties as men. The anti-racist Vibert supported Cheysson's school for housewives, but opposed keeping women in tutelage. Limousin, so vocal a supporter of racial ideology, also thought household schools important, but conceded that women could have political rights.[97] Hence Monin was a conservative on both gender and race, Vibert seemed far more conventional on gender issues than on racial theory, while Limousin was far more flexible on gender than on race.

Cheysson pleaded that he only wished to develop the potential of all women to help distressed working families by learning, for example, how to cure a sick child and how to bring a pure and solid love into the home.[98] The next month the moderate feminist, but pro-suffrage editor of *L'heure de la femme* Lydie Martial (née Anna Carnaud, 1861–1929) asked why not teach men to be fathers and to keep household order? Furthermore, why not allow women full education and initiative in agricultural and colonial enterprises? Cheysson again commented that he would like women to fulfill the familial role first, and then enlarge their place in society, so that their external contacts would then elicit gratitude and recognition.[99]

Vibert raised the Civil Code issue by pointing out that the shame of modern society was the failure of the law to protect women from the abuse or extravagance of drunken husbands. The cleric and philosophy teacher J. A. Clamadieu agreed that there were inequities in adultery provisions in

95 Ibid., 887.

96 On Pelletier, see Scott, *Only Paradoxes*: 125–60; Gordon, *The Integral Feminist*, and "Convergence and conflict," 141–63; Sowerwine, *Sisters or Citizens?* 110–16; and "Woman's Brain, Man's Brain," 289–307.

97 Discussion of Cheysson, "Le Rôle social de la femme," *RIS* 13 (1905): 888–9, 894; for Monin on race, see Discussion of Golberg, *AIIS 1895* 2 (1896): 372–3.

98 "Le Rôle social de la femme," 895.

99 Cheysson, "Le Rôle social de la femme," 44–59, esp. 44, 59; see also Klejman and Rochefort, *L'Egalité*, 174.

the Civil Code and in women being unable to launch paternity suits. This latter issue resulted in the limited reforms of the Rivet law of 1912, which excluded litigation by non-French citizens in the Empire. Clamadieu also believed that educated women would become social treasures.[100]

Moderate sympathy for feminism continued in several periodical articles over the following years. Both a Marxist economics professor from Turin, Achille Loria, and a socialist international lawyer from Modena, Francesco Cosentini, rejected the anthropological theory of brain size, and favoured both civil equality and female suffrage.[101] Loria used Manouvrier's calculations of brain size to show that women's brains had increased in size since the Stone Age, so therefore there was "no ineffaceable inferiority." Cosentini cited several measurements of small brains, such as that of Léon Gambetta, to show that such men were not necessarily less intelligent.[102] Loria predicted that the triumph of feminism would bring a "true civilization based on equality and fraternity." Cosentini endorsed John Stuart Mill's diagnosis that the "social milieu limits women's activity" and that the vote would help women realize their aspirations.[103]

As women's organizations and women members became more vocal, anti-feminists emerged in the next stage of the Société de sociologie de Paris discussion in January 1906. L. Ponsinet of the Châlons prefecture found women's claims too audacious, not so much because of an essential feminine nature, but because each sex has its sociological function for the common interest and solidarity. He, too, depicted women as family educators, benefactors, and moralizing pacifiers of conflict.[104]

The president of the society in 1901–02, the positivist physician Ernest Delbet, was apparently more moderate on the race issue, if one may judge by his hearty welcome for the Haitian lawyer and advocate of racial equality Anténor Firmin. Delbet had avowed that Firmin's participation in a discussion on gold reserves was "proof of the equal rights and fraternity of all races."[105] Now he endorsed Comte's denial of equality of the sexes in favour of complementarity. For Delbet, "irreducible organic differences" made equality an error both in biology and sociology. The feminist John Stuart Mill was "absolute and metaphysical," and subject to

100 Discussion of Cheysson, "Le Rôle social de la femme," (1906) 65; on paternity suits, see Pedersen, *Legislating the French Family*, 138–61.

101 Loria, "Féminisme au point de vue sociologique," 5–17, esp. 16–17; Cosentini, "Féminisme et science positive," 711–35, 809–36, esp. 717, 834.

102 Loria, "Féminisme," 5; Cosentini, "Féminisme," 713–14.

103 Loria, "Féminisme," 16–17; Cosentini, "Féminisme," 717, 834.

104 Ponsinet, "Le Rôle de la femme dans le progrès de la civilisation," 112, 114, 125.

105 Delbet, Discussion of Limousin, "Question monétaire à propos de l'Exposition de 1900," 289.

the harmful influence of Harriet Taylor. Delbet thought Cheysson's household schools appropriate, while women could provide Christian charity for the weak and sick as part of a new moral solidarity. He foresaw at best a slow evolution of women toward some civil and political rights.[106]

Madeleine Pelletier immediately responded that women did not wish to be men's rivals but to develop their activity in society. Le Bon was wrong to argue increasing differentiation of the sexes in civilization, and Comte was a man of system. Mill, far from being metaphysical, was willing to adjust his views to the temperament of individual women. Jeanne de Maguerie added that without political rights one could not hear the voices of women, which were needed to correct the inequities of the Civil Code.[107]

A relational feminist far more conservative than Pelletier and de Maguerie provoked protests from more radical women, as in the aftermath of the Cheysson paper. Augusta Moll-Weiss, an expert on child rearing and founder of a School for Mothers, spoke to the society in 1910 on the woman housekeeper and feminism. Working class mothers could help their husbands by household savings and promotion of better family health, and thus become better feminists. Even wealthier women needed a similar education about their own physiology, nutrition, and child rearing.[108]

Lydie Martial responded again with a plea to educate fathers, and Novicow promoted the right of women to work outside the home. But La Grasserie thought the time for necessary housework implied limited working hours. He concluded French women were ill suited for "integral feminism," though wealthier women could afford to be like the Anglo-American housewife. Worms himself admitted that factory women needed their wage, and without them the economy would suffer.[109]

In the periodicals, the teacher of German language Théodore Joran (1858–?) expressed the most extreme reactionary views. A conservative Catholic and virtually professional anti-feminist, Joran insisted that young single women should have a dowry.[110] The sexes were not equal in any respect, and the feminists' campaign for divorce by mutual consent (resolutions of feminist congresses in 1896 and 1900) was a

106 Delbet, Discussion of Ponsinet and La Grasserie, "Le Rôle social de la femme," 134–7, 139–41.

107 Maguerie, Discussion of Ponsinet and La Grasserie, 139–40.

108 Moll-Weiss, "La Ménagère et le féminisme," 500.

109 Martial, Novicow, La Grasserie, Vibert, and Worms, Discussion of Moll-Weiss, 507, 509–10, 515.

110 On Joran, see Offen, "Depopulation," 662; Lambert's review of "Autour du féminisme," 819.

rebellion against marriage and showed a fear of motherhood. Feminists denigrated the teaching of household science because they disdained to have a household.[111]

A frequent reviewer in the periodical, Emile Chauffard, a law and letters graduate, shared Joran's aversion for feminists, but thought his condemnations too generalized. Chauffard could not accept denial of the legitimate claims of a single woman to earn a living, nor was the right to vote for women any worse than the voting of "debauched males or morons." The following year Chauffard was exasperated with Joran's caricatures that condemned all feminists as "stupid, immoral, and dishonest."[112]

On the whole, however, Chauffard disliked the most well-known radical feminists. He dismissed a pamphlet by the famous crusader for women's suffrage Hubertine Auclert as "intemperate and intolerant" special pleading, despite the admittedly sound argument. This was the year, after all, when Auclert overturned municipal election ballots. It was the mode of expression that alienated him more than the content.[113] He thought Madeleine Pelletier's pamphlet on the struggle for women's rights harmed her cause more than anti-feminist arguments. She would recommend birth control, permit abortion, and masculinize women, with the unintended consequence that men would become indifferent to them. This "absence of sociological knowledge and balance" should alarm "intelligent women" who "should be on guard against the wrong it causes them."[114]

Ghénia Avril de Sainte-Croix (1855–1939) was a founder of the federation of feminist groups, the Conseil national des femmes françaises in 1900–1901, a teacher of "feminism" at the private Collège libre des sciences sociales, and a crusader against the sexual double standard, the international traffic in women, and regulated prostitution, as promoting slavery for women.[115] In 1907 the future urban sociologist and student of north Africa René Maunier (1887–1951) reviewed the historical account of feminism by Mme Avril. Maunier cautioned that feminism disturbed domestic society, increased divorces, and lowered the birth

111 Joran, "Le Féminisme à l'heure actuelle," 324–5, 327–8, 336.

112 Chauffard, Review of Joran, "Au Coeur du féminisme," 218–9; Chauffard, Review of Joran, "La Trouée féministe," 468–9.

113 Chauffard, Review of Auclert, "Le Vote des femmes," 718.

114 Chauffard, Review of Pelletier, "La Femme en lutte pour ses droits," 235–6; Chauffard, Review of Pelletier, "L'Emancipation sexuelle de la femme," 739–40.

115 Rochefort, "The French Feminist Movement and Republicanism," 88; Sowerwine, *Sisters or Citizens?*, 77; McMillan, *France and Women*, 203; on Mme Avril's fight against white slavery and regulated prostitution, see Offen, "Intrepid Crusader," 352–74, and "Madame Ghénia Avril de Sainte-Croix," 239–55; for biography, Offen, "La plus grande féministe," and Camiscioli, *Reproducing the French Race*, 103, 112.

rate.[116] Thus once again the subtext of population concerns was prominent, but upsetting the household was the major issue.

The Société de sociologie repeatedly displayed these polarities. The tireless suffragist, drama critic, and journalist Jane Misme (1865–1935), speaking on the "woman of letters," insisted that changes in women's lifestyles, including their freedom to write, did not threaten morality.[117] The philosopher and critic of the symbolist poets André Barre (1878–19?) strongly condemned any public activity of women outside of happy marriages. For him, a numerous family was far better than producing books. He expressed this reaction to Misme so strongly that even the positivist presiding officer Grimanelli warned him to be polite. Joran, while acknowledging that the public would appreciate any talented writer, thought women's rivalry with male writers inappropriate. At that point, Worms wondered whether women actually write their own literary works.[118]

A few years later, Arthur Bauer, who won an Academy of Moral and Political Sciences prize for an essay on feminism and moral culture, expressed a similarly patronizing attitude toward aspiring career women. Artistic education was fine, but women needed to learn household skills, needlework, and ethics, to cultivate feminine dress and a feminine walk, and to act with a reserved and prudent grace.[119]

There was no overarching conflation in this discourse of race and gender nor were the women who participated in the debates on race the same as those who participated in the debates on gender. Of those who spoke or wrote on both race and gender, Novicow was one of the few consistent progressives, while Manouvrier, a critic of racial theory, backtracked on his earlier enthusiastic feminism. Letourneau was pessimistic on the evolution of "retarded" races, but quite confident in a better future for women, at the price of "masculinizing" their minds. La Grasserie and Monin were examples of moderate conservatives who were adherents of the older paradigms on both issues. The positivist Delbet seemed more tolerant on issues of race than on gender, while the reverse was clearly true of the racial hierarchy advocate Limousin. Even Vibert was not among the most flexible on the gender issue. Women members were clearly divided between the more advanced individualist feminists such as Maguerie and Pelletier, who were more radical than all the men, and relational feminists such as Moll-Weiss, who advocated lifestyles of gender complementarity, as did the majority of male sociologists.

116 René Maunier, Review of Ghénia Avril de Sainte-Croix, "Le Féminisme," 667.

117 Misme, "La Femme de lettres," 30.

118 Barre and Grimanelli, Discussion of Misme, 30–45, esp. 43, 45.

119 Bauer, "Culture morale et féminisme," 288–9, 296, 298.

CONCLUSION

While Worms and his colleagues may have been deficient in the sophisticated ethnological research of Durkheim or Marcel Mauss, they reflected current interest in aptitudes for professions and the nature-nurture controversy as it affected race and gender. Worms and his associates, whether in the Société de sociologie or the Institut international, had no uniform belief in an indelible human nature. The emerging study of sociology could be expected to highlight, as the Durkheimians did, the importance of social milieu rather than biological endowment. Since the anthropological community was already fragmenting over the validity of craniometry, temperate voices such as Manouvrier and Schrader encouraged the sociologists not to attribute too much to inherited brain size.

Still the *Revue* had once given ample space to the anthroposociology of Vacher de Lapouge and allowed questioning of the potential of African-Americans. Worms himself, the anthropologist Papillault, and hobbyist sociologists such as Limousin relentlessly trumpeted European supremacy to "inferior races."

Two noteworthy conclusions on racial issues emerge from study of the non-Durkheimian sociologists' periodicals. The first is that, as among the Durkheimians, a strong faction – represented by the Russian émigrés Novicow and Roberty in the society, as well as the periodical reviewers Lambert and Duprat – denied the sociological usefulness of the old racial hierarchies. While some scholars see the Dreyfus affair as a critical turning point, this opposition was manifested in 1895, well before the Dreyfus affair became the focus of the national press in 1898–99. The second conclusion is that, while racial hierarchy could sometimes justify colonialism, questioning it did not automatically lead to the disavowal of empire. The sociologists seemed reconciled to the French civilizing mission. Vibert clearly supported it for economic reasons and for the prestige of France. Worms's support stemmed from the need for expansion of all healthy societies. The discourse of associationism that rejected non-Europeans as potential citizens was acceptable even if racial hierarchy was questioned. While there seems to be no indication that liberal or republican discourse entailed imperialism in the minds of the sociologists, the cultural acceptability of imperialism was not seen as violating their republican convictions. The right to colonize and dominate triumphed over any conceivable rights of non-Europeans. Nor did they invoke the universalist arguments used by the administrators of French West Africa, as cited by Alice Conklin.

The sociologists admitted a small group of well-known women members, and conducted extensive discussions on the social role of women.

The intersection of race and gender on questions such as brain size or imperial exploitation of women did not seem to affect the discussion about feminism. Most male sociologists were willing to admit, with the exception of the need for childbearing, that there was no fixed feminine nature. The most militant and individualist feminists, such as the crusaders and pamphleteers for political rights Madeleine Pelletier and Hubertine Auclert, elicited little sympathy. However, most participants in the discussion were sympathetic toward reform of the Civil Code and a racial hierarchy supporter such as Limousin supported political rights for women. The critical stumbling block was not the educability of women but rather male notions of social stability.[120] In this view, women should be subordinate in marriage and retain responsibilities for the household.

A coterie of positivist traditionalists (such as Delbet and Grimanelli, who played important official roles in the society) remained, as well as a few virulent anti-feminists such as Joran and Barre. For them, scientific arguments about women's nature were an auxiliary tool to assert that women should never challenge male social dominance and should remain in complementary roles. In the crucial suffrage decision, when the Senate rejected the vote for women in 1922, there were ready-made arguments that republican citizenship for women would destabilize society and distract women from their natural functions.[121]

Considering Durkheim's own views on gender issues, one could conclude that many of the non-Durkheimians were hardly less progressive. However, the Durkheimians were less likely to act as a minority did among the non-Durkheimians – supporting the old anthropological paradigm as a matter of principle. The range of non-Durkheimian views shows the weakness of proponents of indelible nature as well as the social obstacles to admitting women to full citizenship. Neither gender stereotyping, nor the racial paradigm, nor the enthusiasm for empire would disappear in the years before World War I. The enshrinement of society and culture as all-important explanatory entities in the social sciences would gain momentum in the interwar years.[122] The sharp decline in racial stereotyping, at least in the scholarly community, along with the evident triumph for women's suffrage, had to await the aftermath of World War II and, in some cases, the era of decolonization.

120 Offen, "Depopulation," and Surkis, *Sexing the Citizen*, 113–14.

121 Hause, "More Minerva than Mars," 97–104; McMillan, *France and Women*: 217–18, 229–30.

122 Conklin, "The New 'Ethnology'," 29–46; Paligot, *La République raciale*, 279–315.

8

Reorientation of Institutions

THE NEW ETHNOGRAPHY: DELAFOSSE AND RIVET

THE PERIODICALS IN PSYCHOLOGY and sociology demonstrate an early twentieth-century trend away from the assumptions of physical anthropology, though neither psychologists nor sociologists completely disowned hierarchies of race and gender. The world of ethnography and anthropology before 1914 saw a wholesale rearrangement of institutions that manifested a questioning, though not abandonment, of the prevailing evolutionary schema of "retarded" races or cultures. Imperial discourse would still flourish in almost all quarters, while hard-line mainstream anthropologists persisted in deploying physical measurement. David Hoyt has called this relative shift, predating the extensive fieldwork of Bronislaw Malinowski and Alfred Radcliffe-Brown, the "reanimation of the primitive" – the perception that savages were not dead cultures needing to advance through historical stages, but vital cultures capable of agency and needing a synchronic, functionalist analysis.[1] The shift, while incomplete, had several catalysts. The critique of cranial and corporeal measurements from within the anthropologists' own ranks and the lack of international standards had opened the way for a more truly ethnographic perspective. The passing of an older generation of "scientific materialist" anthropologists also produced opportunities for different approaches.[2] The context of the Dreyfus affair brought the Durkheimians together in a common front against the excesses of racial hierarchy and indeed they found no common definition of race.[3] For the Durkheimians

1 Hoyt, "The Reanimation of the Primitive," 331–54, esp. 340–2 on van Gennep.

2 Zerilli, *Il Lato oscuro*, 39, 48–9.

3 Mucchielli, "Sociologie versus anthropologie raciale," 77–95.

explanation of complex social phenomena by simple or more rudimentary forms did not necessarily imply the inferiority of "archaic" societies.[4]

The new trend is perhaps best represented by the determined initiative of folklorist Arnold van Gennep to separate ethnography from physical anthropology while avoiding the literary bent and eccentricities of the previous Société d'ethnographie. The independent periodical *L'Anthropologie*, edited by René Verneau and the paleontologist Marcellin Boule, also showed more openness to the nuanced insights of colonial agents and administrators in Africa, who contributed to the reorientation.[5] Influential individuals who orchestrated this reassessment included the west African colonial administrator and African languages teacher at the Ecole des langues orientales Maurice Delafosse (1870–1926), and the military physician and Americanist Paul Rivet (1876–1958). Both were trained in the techniques of physical anthropology at the Muséum as pupils of Ernest-Théodore Hamy, but both came to value ethnographic and linguistic analysis.

Delafosse reassessed African culture favourably and supported a migrationist and diffusionist approach to cultural development. In 1893, even before joining the colonial administration, he reported the frank, hospitable, and democratic sentiments among the Ivory Coast Agni people.[6] Later he found that the "best revelation of intellectual superiority" among the Vaï of Liberia and Sierra Leone was their "being refractory to European civilization."[7]

But Delafosse's comments on African development show just how incomplete and halting were his revaluations. He did not surrender interest in nature to nurture, or give up colonial objectives based on the need for African cultural development. Delafosse retained the framework of white, yellow, and black "great races," believed physical anthropology useful for ethnic classification, and looked unfavourably on inhabitants of equatorial Africa. He also believed that Europeans initiated the great cultural productions of Africa.[8]

His most famous parallel appeared in a brochure edited by L. de Morrens (probably a pseudonym for the colonial lobbyist Léopold de Saussure, whose family had been seigneurs in the Swiss village of Morrens) about the allegedly civilized black Liberians and the indigenous, allegedly "savage" tribe of the Baule. He remarked that he wished

4 Karady, "Durkheim et les débuts de l'ethnologie universitaire," 24–32.

5 Sibeud, *Une Science impériale pour l'Afrique?*, 17, 140–4.

6 Delafosse, "Les Agni," 402–24.

7 Delafosse, "Les Vaï," 133.

8 Sibeud, *Une Science imperiale*, and Amselle et Sibeud, *Maurice Delafosse*; Conklin, *A Mission to Civilize*; Paligot, *La République raciale*, 225–7, 239.

to be "neither Negrophile nor Negrophobe, but simply a Negrologue," a student of Africans. The history of Liberia demonstrated that Africans could progress toward European civilization, but the Liberians lacked the moral sense of the Baule villagers and they appeared to be victims of arrested development.[9] The villagers showed great dignity, treated women well, and were not habitual thieves or drunkards.[10] Europeans should therefore realize, according to Delafosse, that "blacks are not now apt to take on European civilization, but are not inferior either." Criticizing the naval physician Laurent-Jean-Baptiste Bérenger-Féraud, he concluded, no people deserved to be called savages or inferior because "they don't wear shoes or corsets, are indifferent to stock prices, and because we haven't been able to tame or fashion them as we wish."[11] Nevertheless, given present European superiority, the Africans could not be expected to catch up in a day or even in a century, and the special genius of each race did not mean equality of all.[12] This latter perspective is only slightly more advanced than that of the evolutionist Charles Letourneau.

The affinity of Delafosse for the imperial discourse of associationism was obvious in his review of the Indochinese administrator and anthropology society member Jules Harmand's classic work *Domination and Colonization* (1910). Delafosse approved Harmand's stipulation that self-government was inappropriate for "subjects of foreign races." He also concurred with Harmand that "one cannot elevate the yellow and black races in the social hierarchy except by a certain acceleration of their march and not by a deviation from the ancestral road they have to tread." So while Delafosse had inclinations toward cultural relativism, he remained in the ambit of imperial discourse and gradualist improvement for the colonized.[13] In 1920 he taught French West African customs at the training school for colonial administrators, the Ecole coloniale. He wrote in the bulletin of the imperial lobby group, the Comité de l'Afrique française, of the need for profound knowledge of native societies, which had a right to conserve their own civilization and provide for their needs according to their own temperament and genius.[14]

9 Delafosse, "Les Libériens et les Baoulés," 97–122, 129–40, esp. 102, 105–7; see editor's comment, 127, which suggests he wrote *Psychologie de la colonisation française*, a well-known work of Léopold de Saussure.

10 Ibid., 110–12, 130.

11 Ibid., 137, 140; Delafosse also maintained his conceptions of African incapacity to Europeanize in a 1912 review; see Zerilli, "Maurice Delafosse," 149.

12 Paligot, *La République raciale*, 244–5.

13 Delafosse, Review of Jules Harmand, "Domination et colonisation," 65–6.

14 Wilder, "Colonial Ethnology and Political Rationality," 229–31.

Delafosse's great accomplishment, compared to nineteenth-century mainstream anthropologists, was to banish the idea of irremediable racial inferiority as such, and to laud the creativity and future potential of Africans.[15] He was more positive than Letourneau, though he retained remnants of a stages theory of development. After World War I, Delafosse made clear that he would not justify self-determination for the colonies on the basis of a few Europeanized Africans.[16]

L'Anthropologie also became hospitable to cultural and social approaches, including reviews or articles by Durkheimians such as the Mexicanist and Arctic ethnologist Henri Beuchat and the sociologist of religion and prehistorian Henri Hubert. The antiquities museum director Salomon Reinach critically discussed Durkheim's views on incest prohibition, totemism, and exogamy.[17] Muséum professor and editor René Verneau commented on the work of Henri Hubert and Marcel Mauss on the function of sacrifice as expiation, regeneration, and request.[18] A certain "J.-P." reviewed Mauss's famous paper on seasonal variations in Inuit communities and later on the general theory of magic.[19]

Early on, the periodical editors kept their distance from the anthropology society. Emile Cartailhac questioned the Godard prize award decision of the central committee of the Société d'anthropologie de Paris that annulled a decision of its own commission and implied that, unique among scholarly societies, this was not a "liberal" organization.[20] Iberian language expert Julien Vinson had even opposed allowing Verneau, an editor of *L'Anthropologie* since 1894 and also president of the society in 1902, to teach a course in the Ecole d'anthropologie in 1905.[21] In 1907, when Adrien de Mortillet opposed municipal subsidies for an anthropology chair, Verneau lamented what he suspected

15 Delafosse, *Les Noirs de l'Afrique*, 156–60.

16 Paligot, *La République raciale*, 261.

17 Reinach, "La Prohibition de l'inceste," 59–70 and (1902) 655–69; see Hoyt on "Totemism, Ethnography, and the Crisis of Liberal Politics."

18 Verneau, Review of Henri Hubert and Marcel Mauss, "Essai sur la nature et la fonction du sacrifice," 106–7.

19 J.-P., Review of Marcel Mauss and Henri Beuchat, "Essai sur les variations saisonnières des sociétés Eskimos," *L'Anthropologie* 17 (1906): 711–14; Review of Mauss, "Essai d'une théorie générale de la magie," 19 (1908): 360–2.

20 Cartailhac in *L'Anthropologie* 1 (1890): 150; the Archives de la Société d'anthropologie, Comité central at Musée de l'Homme indicate that as early as 1870 the question of allowing all members to elect the committee was discussed, but rejected on the grounds that it would be unseemly to discuss the merits of candidates at a public meeting and that provincial members would not be well enough informed about the scientific merit of the candidates.

21 Zerilli, *Il Lato oscuro*, 90.

was a personal attack on him, since he had lectured at popularized city-sponsored lectures in anthropology since 1895.[22]

Paul Rivet (1876–1958), a military physician and from 1909 an assistant to René Verneau at the Muséum, also contributed to *L'Anthropologie* in this era.[23] In 1925 he would help found the Institut d'ethnologie with Marcel Mauss and Lucien Lévy-Bruhl. The institute would re-orient the emphasis of anthropology (or ethnology, as he preferred to call it) from physical anthropology to ethnography, archaeology, and linguistics. He would also revitalize the Musée d'ethnographie of the Trocadéro and in 1938 inaugurate it as the Musée de l'Homme, a major teaching and research centre as well as a place for exhibits.[24]

In Ecuador for a military geodesic mission, Rivet was contemptuous of the customs and lifestyle of the native South Americans of the Riobamba region, but he thought their faults "logical and necessary consequences of miseries endured and not the sign of a primitively bad nature." Pity was the appropriate response. Then, too, Europeans could "with patience and gentleness gradually elevate this stationary and immobile race, to bring it out of its purely animal life, in a word to make of it a race of men." Geography, climate, and history were all responsible for their degradation rather than racial heredity. The Spanish had brought alcoholism and smallpox at a time when these peoples were ready for progress. Instead they emerged like beaten children.[25]

Having spent a month among the Colorado people of the Ecuador plateau, Rivet concluded that contact with Europeans was far from beneficial to the Native Americans. "Truly one must not study too much in detail the work of civilization, but only its general effects. Otherwise one would be led to doubt its beneficent influence, or even progress."[26] His papers on the Jibaros of Ecuador praised their refusal to accommodate to European conquest while admitting they could not be civilized. Less impressive for Rivet was the syncretism of their Christian and indigenous religious practice. As with those around Riobamba, he tended to be critical of Native American religions laced with what he considered superstition and only superficial adaptation to the civilized spirituality of Christian doctrines.[27]

22 Verneau in *L'Anthropologie* 18 (1907): 240–1.

23 On Rivet see Laurière, *Paul Rivet.*

24 Conklin, "Civil Society, Science, and Empire," 255–90.

25 Rivet, "Etude sur les Indiens de la région de Riobamba," 78–9; Laurière, *Paul Rivet*, 148, 151.

26 Rivet, "Les Indiens Colorados," 208.

27 Rivet, "Le Christianisme et les Indiens," 81, 85; "Les Indiens Jibaros," 343; Laurière, *Paul Rivet*, 50, 68–71, 123, 128–9, 136, 139.

However, even Rivet, a member of the new Société des américanistes (1895) from 1907, was not above making racial distinctions. At this early stage in his career, Rivet was as much committed to cranial measurements, notably of the Native Americans of Ecuador, as to the study of language and customs.[28] As someone believing in European cultural superiority, Rivet remained ambivalent on the concept of racial hierarchy. But he later added the proviso that even degenerate non-European races could always progress in the future. In 1904, however, he predicted the rapid disappearance of native South Americans after European commerce opened up the forest, as they could not survive contact with a "stronger, more active, more intelligent race."[29]

The major re-evaluation of racial hierarchy was evident, however, in his refusal to adopt the conventional wisdom in craniology that a "prognathic" face (protruding jaws and receding foreheads), any more than cephalic index or cranial volume, signified cultural inferiority or a low rank in the hierarchy. He agreed that the facial angle could differentiate humans and primates, or even indicate the degree of mixing of races, but it could not justify serial classification of races, since prognathism varied much within each population. Furthermore, the Australians and Fuegians were not distant from the Japanese and the French Auvergnats in the relevant facial measurements. Bushmen yielded similar measurements to Bavarians, with no corresponding correlation to intellectual evolution.[30]

In response to the notorious questionnaire of the anthropologists on racial mixing, in 1912 he approved of crossing white and Native American races, but not white and black.[31] Around the same time he was also ready to concede that Amazonian "savages" were not a static people, but had their own "civilization" and contributed much to Andean civilization.[32]

Rivet later wished to correlate physical measurements with linguistic evidence and observations of religion, daily life, and material culture. A member of the Société d'anthropologie since 1902 and active from 1907, Rivet left the Society in 1910, probably because of the commitment of the anthropologists to continue a strongly physicalist approach to their discipline and because of disputes about the questionnaire on

28 For a fuller listing of Rivet's publications in the *Journal de la Société des américanistes,* founded in 1895 by Hamy, see Edison, "'Red-Skins' and White Legends," 10–11 and the bibliography in Laurière, *Paul Rivet.*

29 Rivet, "Les Indiens de Mallasquer," 152; Laurière, *Paul Rivet,* 33, 144.

30 Rivet, "Recherches sur le prognathisme," *L'Anthropologie* 21 (1910): 505–18, 637–69, as cited in Zerilli, *Il Lato oscuro,* 65–72; see also part 1 of the article in *L'Anthropologie* 20 (1910): 35–49, 175–87.

31 Zerilli, *Il Lato oscuro,* 66; Wartelle, "La Société d'anthropologie," 159; Paligot, *La République raciale,* 283, 302–4; Laurière, *Paul Rivet,* 72–90, 105.

32 Ibid., 265.

Métis formulated and circulated from 1908 to 1910. The society's Adrien de Mortillet alleged that Verneau wished to control the society in 1910. After Verneau's resignation, Mortillet attacked the psychologist Henri Piéron as a "liegeman" of Verneau, and a lab technician had to separate the two after a brief fistfight.[33]

Rivet helped form the Institut français d'anthropologie in 1911, a coalition of Muséum anthropologists and Durkheimians with politically progressive inclinations and a determination to highlight linguistic, ethnographic, and archaeological methods, though in practice the organization was more ecumenical. By the time of the foundation of the Institut d'ethnologie in 1925, Rivet, despite teaching courses in anthropometry, was closer to the Durkheimians than to his former colleagues in the Société d'anthropologie. He disagreed with the Durkheimians' overemphasis on Australians. Like them, though, he was now inclined to privilege non-craniometric methods. Indeed he argued in the 1920s that linguistic data for ethnic diagnosis were more certain than craniometry.[34]

By 1928 Rivet had achieved a prominent position in the anthropological community. After a contentious campaign, he secured appointment as Muséum professor of anthropology as well as director of the Musée d'ethnographie of the Trocadéro. The old stalwarts of the Société d'anthropologie had opposed him, with anatomist Raoul Anthony and Muséum paleontologist Marcellin Boule backing physical anthropologist Henri Vallois. In 1936 Rivet would change the name of the Muséum anthropology chair to "ethnology of modern man and fossilized man."

In the interwar and post-World War II period, Rivet would argue that the technical skill of Native Americans showed their inventive genius, and he would enumerate their many contributions to European civilization.[35] He also would tell Protestant missionaries in 1931 that though European technical, moral, and social progress was admirable, Australians could not be considered "primitive"; rather they were a complex civilization that had evolved in a different direction. Yet his remarks that the distance between an Einstein and "primitives" was greater than between "primitives" and anthropoid apes recall nineteenth-century clichés.[36]

In 1936 Rivet edited an encyclopedia volume that included articles justifying racial hierarchy, but he would be notably active in the anti-fascist vigilance committee of intellectuals from 1934 to 1937. He would also enthusiastically co-operate with Célestin Bouglé on the anti-racist *Races et racisme* periodical published from 1937 to 1939. The brochures of this

33 Laurière, *Paul Rivet*, 44; Wartelle, "La Société d'anthropologie," 160.
34 Laurière, *Paul Rivet*, 240, 279.
35 Ibid., 273–5, 360–8.
36 Ibid., 654–5.

group frontally attacked ideas of racial purity and superiority as well as the specific Nazi classifications of Aryans and Jews. While never discarding race as a biological division, Rivet identified with those attacking the stability of types and the determination of mind and character by race.

During the Vichy era Rivet lost his position at the Muséum because of his sympathy for the Resistance circle among his pupils and colleagues, and because of controversial open letters to Marshal Pétain critical of his regime. He took refuge in Colombia, where he endorsed the Free French, founded and taught at the Colombia National Institute of Ethnology, and also continued his ethnographic research. He continued, without romanticizing the past, to highlight the position of Native Americans as inventors and actors in the South American context and in their contributions to European and global culture. After the war he was able to resume all his previous positions in Paris until his official retirement in 1949.[37] The latter part of Rivet's career would be complicated by his support for French Algeria and opposition to Algerian nationalism due partly to his friendship with the Americanist ethnologist Jacques Soustelle, governor-general in Algeria in 1955–6. However, Rivet remained a strong opponent of racism while paradoxically clinging to respect for the "civilizing mission" of France.[38]

ARNOLD VAN GENNEP

As the old Société d'ethnographie entered a moribund phase from 1902 to 1913, the journalist-folklorist and future teacher of comparative ethnology at Neuchâtel, Arnold van Gennep (1873–1957), criticized the state of French ethnography in a series of articles beginning in 1905 in the literary periodical *Mercure de France*.[39] In 1908, during his brief unhappy membership of the Société d'anthropologie de Paris (1905–09) he founded a new periodical, *Revue des études sociologiques et ethnographiques* (later *Revue d'ethnographie et de sociologie*).[40] In 1910 he and Delafosse would launch the new Institut ethnographique international, which was active until 1914, when it had over 200 members, including a strong

37 Paligot, *Races, racisme et antiracisme dans les années 1930*, 40–5, 49; Laurière, *Paul Rivet*, 487, 494–7, 527, 534, 580–1, 595–7; for text of the letters to Pétain, 665–70; for *Races et racisme*, 502–16.

38 Ibid., 613–17.

39 Zerilli, *Il Lato oscuro*, 100–1.

40 On van Gennep, see Centlivres and Hainard, eds, *Les Rites de passage aujourd'hui*; Fabre, "Le *Manuel de folklore français* d'Arnold van Gennep," 641–75; Centlivres and Vaucher, "Les Tribulations d'un ethnographe en Suisse," 89–101; Sibeud, "Un Ethnographe face à la colonisation," 79–104.

colonial contingent.[41] The Institut had strong support from some of the same bodies that supported the older Société d'ethnographie, such as the Langues orientales school (the first president, after the mining engineer Jacques de Morgan in the organizational period, was Henri Cordier, an expert on Chinese and Indochinese documents). Also prominent were lecturers from the Collège de France, and a diverse array of non-specialist geographers, mining engineers, publishers, artists, and anthropologists such as Joseph Deniker.[42] Durkheimians such as the Americanist Henri Beuchat joined the editorial board. Eventually an important contingent of ethnographer-administrators from Africa also joined.[43]

The program of the Institut ethnographique international aspired to be "descriptive and theoretical, a study of life in society" of men from all times and all countries, amalgamating the "savage" and civilized in a fashion that the old Société d'ethnographie had avoided. However, the organizers deliberately excluded what they called "anthropology proper, or anatomical study of human varieties, as not in our framework except insofar as it promises to define the relationship that can exist between races of men and their civilizations or the influence of diverse civilizations on each other." Contributors to the journal were far more inventive, problem-oriented, centred on material culture, and, most of all, interested in fieldwork techniques than the old Société d'ethnographie.[44] Van Gennep himself had been contemptuous of Rosny's society.[45] A sign of his attack on racism appeared in a review of a lecture by the famous anthropologist Franz Boas, where van Gennep asserted, "it is impossible to pretend one race is physically superior to others." Moreover, among various civilizations, the "forms of the beginning are often more complex and forms called more recent are often the most simple."[46] Hence, ethnography should not discriminate among cultures, though it might still have a role in providing useful colonial information.

41 Sibeud, "Un ethnographe," 80.

42 Sibeud, *Une Science impériale*, 160.

43 Sibeud, "Metamorphosis of Ethnology," 102.

44 *Revue des etudes ethnographiques et sociologiques* 1 (1908), Prospectus; van Gennep's Institute included disparate individuals such as the Durkheimian Henri Hubert and the disciple of the sociologist Frédéric Le Play, Louis Marin, who taught ethnography at the private Collège libre des sciences sociales.

45 Sibeud, *Une Science impériale*, 155, citing his 15 February 1905 article in the *Mercure de France* on "Ethnographie, folklore"; the Société d'anthropologie de Paris in the person of Near Eastern archaeologist and Collège de France lecturer René Dussaud gave only lukewarm reviews in Review of Van Gennep, "Rites de passage," 141–2 and Review of Van Gennep, "Mythes et légendes de l'Australie," 222–3.

46 Van Gennep, Review of Boas, "Anthropology," 289–90.

Yet there was no complete break with the study of races. Even van Gennep, in his 1914 commentary on Algeria, found an "arrested development" of native races, because he thought that after young people reached adolescence, intellectual regression occurred.[47] He denounced French settlers who refused the political assimilation of educated Algerians. Still he accepted the Kabyles, according to the old mythology, as an indigenous elite compared to the Arabs. His perceptions of the Spanish and Jews in Algeria were unfriendly, and he expressed irritation at the "indigenous mentality" that created obstacles to his interviews for the study of native crafts. Nor was he averse to collaborating with peaceful, reformed, non-exploitative varieties of colonization.[48]

In its Neuchâtel congress of 1914 on ethnology and ethnography, the new institute planned to highlight psychology of religion, sociological ethnography, and folklore, in addition to ancient civilizations. The French delegation included Marcel Mauss of the Ecole pratique des hautes études and René Verneau of the Muséum, who was piqued that the name ethnology (in the old sense of physical description of peoples or a science of races) had been used without really being on the program.[49]

While Verneau lamented the potential scattering of effort and lack of resources, van Gennep, in an article in the *Mercure de France* after the congress, hoped that it would indicate the definitive separation of the "sciences of man" from the "disciplines that until now have succeeded in stifling them" and insisted that ethnography would have received public recognition sooner if "in its moment of rebirth it was not strangled by the anthropologists." Paul Broca had unjustifiably annexed ethnography, while it should have had little to do with anthropology. Even Marcel Mauss, the arch-advocate of a revivified ethnography, was suspicious, according to Verneau, of van Gennep's separatism.[50]

On the whole, then, van Gennep's venture was a signal demonstration of the effort to replace undue emphasis on nature rather than nurture. He was as caustic as Léon de Rosny about craniology. But while his efforts did not result in a complete disavowal of anthropology or a complete break with racial ideology or imperialism, he prepared the way for others to transform ethnology in the Institut d'ethnologie of the Sorbonne.

47 Sibeud, *Une Science impériale*, 242–3, and Paligot, *La République raciale*, 208.

48 Sibeud, "Un Ethnographe," 83–4, 97–9.

49 See the plans in *L'Anthropologie* 24 (1913); for an overview see Zerilli's article "Etnografia e etnologia al congresso di Arnold van Gennep," 143–52.

50 Van Gennep, "La Signification du 1er Congrès d'ethnographie, 322–4, cited also in Zerilli, *Il Lato oscuro*, 103, 105.

PERSISTENCE OF THE SOCIETÉ D'ANTHROPOLOGIE

Georges Papillault, Manouvrier's assistant in the anthropology lab and sociology lecturer at the Ecole d'anthropologie, represented the old mainstream trend of physical measurements and racial hierarchy. In some respects Papillault was the anti-Manouvrier. In the Paris Société de sociologie in 1902 Papillault commented, with reference to Raoul de La Grasserie's paper on race and milieu, that race should not be minimized. He insisted that specific organic characteristics, such as the weight of the brain, structure of face, muscles, and erect posture influenced the evolutionary position and rank of races. The most evolved race correlated with the most complex society, as Letourneau had asserted. He was willing to admit that education could influence individuals who would outpace the average European. In a good mixed society, each race would have the proportion of members corresponding to its organic value.[51]

Papillault would continue to argue for a "racial sociology" with permanent stability of racial types. He was active in the French colonial congresses of the first decade of the century. In 1912 he still questioned whether human groups were separate species.[52] Papillault particularly wished to blunt the opposition to Lombroso's criminal anthropology and cited Galton's studies on heredity to minimize the possibility of improvement through education. Certainly unfavourable milieux and economic motivation affected actions, but some predispositions were undeniably important. Certain types of "degenerates" with physically observable markings were more likely to be criminal than others. In an obituary for Galton, he noted the plea for eugenic duties to humanity to make well-selected marriages to combat degeneration.[53] Papillault agreed that inequality was indispensable to progress and opposed a "childish Lamarckism."[54] In this case neo-Lamarckianism, which could have ambivalent political connotations, was considered a left-wing aberration.

Despite his physicalism, Papillault as a sociology teacher had a sense of the wide scope of anthropology. He lambasted Swiss authors for wanting to restrict anthropology to the physical because they feared a naturalistic anthropology might imperil the existence of the soul. He claimed Edward Burnett Tylor, James Frazer, and the society's own Letourneau as

51 La Grasserie, "La Race et le milieu," 387–90.

52 Sibeud, *Une science impériale*, 71, and Zerilli, *Il Lato oscuro*, 111.

53 Papillault, "Sur Quelques erreurs de méthode en criminologie," esp. 332; see also "Galton et la biosociologie," 61, 64.

54 Papillault, Review of Lucien Cuénot, "La Genèse d'une espèce animale," 154.

anthropologists. The study of customs, sociology, and political economy could all be integrated under its jurisdiction.[55]

He thought political economy was the model social science and believed that, to confront problems such as urban crime, one needed to study the morphology of races and degeneracy as well. The efforts of Gobineau and Vacher de Lapouge were premature, and criminal anthropology abused measurements, but psychological tests of attention and prolonged examination of a subject's work might be helpful.[56] He also hoped to look at the family milieu, the moral and economic situation, the religious and professional milieu of parents, and the incidence of alcoholism, prostitution, and vagabondage. Punishment for crime needed adjustment to individual characteristics. As was fitting for a future member of the French Eugenics Society (founded in 1912), he felt that Galton's Eugenic Lab had already achieved impressive results, and further inquiry would be better than literary and sentimental declamations.[57]

Just at this time Manouvrier commented on the cases of a few young criminals in an impressionistic way and denied the importance of degeneracy when milieu itself provided enough causes for prostitution. Manouvrier was still adamant that criminal behaviour was a violation of social laws, not a matter of anatomy or heredity. Criminal tendencies themselves were normal and present in everyone, but the wealthy often satisfied their needs within the law, even if their actions were morally illegitimate.[58]

Manouvrier had long argued that heredity left greater latitude than generally acknowledged in the genesis of crime.[59] A prominent jaw might only signify a muscular person, not brutality. According to both Henri Joly and Gabriel Tarde there was no general tendency to theft or murder in prehuman or prehistoric ancestors, in contradiction to Lombroso's equation of savagery and crime. One might as well have said that sulphuric acid has criminal tendencies, since our atavism goes back to chemicals in our bodies.[60]

According to Manouvrier, troubled individuals may in fact have tendencies toward weak inhibitions – a morbid heredity. But no general pronouncement about savages could be made when "European armies rival each other with ardour for machine-gunning savages in order to

55 Papillault, "L'Anthropologie est-elle une science unique?" 117–32.

56 Papillault, "Bio-sociologie," 1–19; Zerilli, *Il Lato oscuro*, 89–90.

57 On legal and political implications of the questions of métissage, see Saada, "Race and Sociological Reason in the Republic, 361–91; for Papillault in the Eugenics Society, see Schneider, *Quality and Quantity*, 89.

58 Manouvrier, "Quelques cas de criminalité juvenile," 428.

59 Manouvrier, "La Genèse normale du crime," 405–58.

60 Ibid., 419.

share their country among themselves."[61] All men had continuity with animal nature and had the potential to kill other creatures. The inculcation of hunting by tradition, imitation, and repetition was as important as some innate impulse to hunt. Men are no more born criminals than dogs are born swimmers.[62] Abnormal characteristics among prisoners may be coincidental, not signs of criminality, and those cerebrally inferior might only show that the smartest criminals do not end up in prison.[63]

Papillault's response to Manouvrier's criminal anthropology railed against unrealistic theories of reformation of criminals that stemmed from the careerism of reformers with jobs in sanatoriums and asylums.[64] No conclusions could be based on just a few cases. Both a bad environment and the personality of a "hereditary dangerous pervert" might be responsible for crime. Anthropometry might help since signs of degeneracy might be apparent in measurements of the biological and social heredity of the subject.[65] Hence Manouvrier and Papillault clearly represented the ongoing and as yet unresolved feud in criminological anthropology between a predominantly hereditarian view and openness toward the strength of the social milieu.

The question of the fertility of racial crossing also produced key lines of demarcation in commitments to rigid racial hierarchy. From 1906 Georges Hervé (1855–1932), ethnology professor at the Ecole and president of the society in 1898, was bent on confirming a worldview akin to Broca's: that races were not fully fertile after interbreeding. For him Métis exhibited physical and moral decline, a view of Gobineau's. Without infusion of black or white blood, he claimed third-generation mulattoes would be sterile. Coexistence, not fusion, of elements took place, with less fertile and less vigorous offspring. He could not recommend racial mixing for the masses. His examples were taken almost entirely from the alleged observations of black-white crossing in the United States.[66] Sigismond Zaborowski, the new ethnography professor, pointed out the apparently low fertility in Australian-English Métis. Paul Rivet criticized these contentions because his South American experience had led him to believe in different types of crossing for different races. Latin American mestizos were not declining and might even be more robust than the mother races. He would soon argue that racial crossing might

61 Ibid., 428.

62 Ibid., 441, 443.

63 Ibid., 449–50.

64 Papillault, "Condition d'une enquête scientifique sur les criminels," 41–56, esp. 45.

65 Ibid., 47, 55.

66 Hervé, "Le Croisement des races aux Etats-Unis," 337–58; see discussion in Zerilli, "Il Dibattito sul meticciato," 237–73.

even be a factor in progressive evolution. The Sorbonne physiologist Louis Lapicque also thought social or political factors were operative in Australia, so no racial incompatibility should be assumed.[67]

In 1907 Papillault and Hervé established a special Commission for the Study of Ethnic Crossing (which included the anatomist and Ecole d'anthropologie professor Raoul Anthony, the physician Emile Atgier, the prehistorian Marcel Baudouin, Medical Faculty zoologist Raphaël Blanchard, Muséum librarian Joseph Deniker, the explorer Oscar Schmidt, the parasitologist Lucien Semichon, the physician Henri Weisgerber, Sigismond Zaborowski, and, before his resignation, Paul Rivet). The commission questionnaire, prepared in 1908 by Papillault, sought to determine whether the great human groups were in fact separate species by probing the health and fertility of Métis. Those who believed that races were not mutually fertile now justified their viewpoint not from Galton but from Mendel – by the reappearance of pure types.[68] Some hypothesized that one race would become dominant at that point. Papillault postulated the least fertility in races farthest apart. The questionnaire included the classification of Hindus and Semites as separate races from Europeans.[69] The commission hoped to explore both physical and moral/intellectual qualities of Métis, including criminal tendencies.

They found many Africans unaware of their remote ancestry or unwilling to answer questions asked by the administration. The entire inquiry had the problem of how to define pure racial entities and whether any meaningful data could be produced from the definition. Some Moors and Peuls (Foulah) were already products of crossing and could not be considered pure races. The published responses from French West Africa showed the small number of Métis in the census, with some reluctant to acknowledge non-European parents and most still young children about whom there could be little intellectual or moral diagnosis. Some administrators favoured assimilation as soon as possible, while others cited Le Bon on the poor moral character of Métis. More conclusive replies to the questionnaire never appeared. In the 1930s equivalent surveys could have consequences for the granting of French citizenship and the treatment of abandoned children.[70]

In the prewar decade Papillault, Hervé, and Zaborowski were still concerned about the primacy of physical anthropology. They had a vision

67 Zerilli, "Il Dibattito," 247, 249–50, 253.

68 "Société d'anthropologie de Paris, Enquête sur les croisements ethniques," 337–8; for replies to the questionnaire, 340–407.

69 Ibid., 345.

70 Saada, "Race and Sociological Reason."

of racial hierarchy that had been only slightly amended since the days of Paul Broca. They prefigured the continuing issues that would come back to haunt opponents of racial hierarchy in the 1930s and 1940s. For them indelible inherited identities left limited flexibility for improvement.

THE INSTITUT FRANÇAIS D'ANTHROPOLOGIE

While the results of the survey on Métis were inconclusive, the disillusionment of Rivet, as well as of Verneau, Lapicque, and the biologist Etienne Rabaud, seems to have been a key factor in the foundation of a new rival Institut français d'anthropologie.[71] This relatively small group appeared at the Muséum in 1911 with an imposed limit of only fifty regular academically elite members.[72] The mission of the new institute was to "bring together specialists of diverse sciences whose union constitutes the science of man." A key patron was the geographer and Academy of Sciences member, Prince Roland Bonaparte. There was a heavy concentration of professors, from the Sorbonne (including Emile Durkheim, philosopher Lucien Lévy-Bruhl, and Lapicque), the Muséum (René Verneau, his assistant Rivet, paleontologist and editor of *L'Anthropologie* Marcellin Boule, and librarian Joseph Deniker), the Collège de France (Sanskrit scholar and Indochina expert Louis Finot), and the Ecole pratique des hautes études (Durkheim's nephew Marcel Mauss). The first president was the historian of religion and Saint-Germain antiquities curator Salomon Reinach: Boule became second president in 1914; and Verneau assumed the presidency in 1922. Some ethnographer-administrators from the Empire attended sessions of this group.[73]

In its introductory session in the Muséum anthropology laboratory, Salomon Reinach hoped for a "happy alliance of the method of the natural sciences and historical method."[74] Potentially the institute seemed to foster an alliance of the best work in both physical anthropology and ethnography – open to all fields of anthropology.[75] However, the dissenters in the debate on racial mixing in the Société

71 Wartelle, "La Société d'anthropologie," 160; Cohen, "Sur l'ethnologie en France, 85–96; Zerilli, *Il Lato oscuro*, 95–6.

72 Ibid.

73 Sibeud, "Metamorphosis of Ethnology," 106–8.

74 Reinach, *Institut français d'anthropologie. Compte-rendu des séances*, 8.

75 Supplement to *L'Anthropologie* 22 (1911): 10–43.

d'anthropologie – Lapicque, Rivet, and the biologist Etienne Rabaud – all seemed to gravitate to the new institute.[76]

But, interestingly enough, this group did not suddenly all become experts on culture. Even the Institut français d'anthropologie remained interested in measurement and illustrated the balance of interests between cultural and hereditary physical factors. Lapicque's papers within the institute were still devoted to the eye and Asian cranial weight and volume. Like Verneau, he doubted Boas's views on the influence of the milieu on cranial type. He was more convinced of the significance of Mendelian laws for human heredity than Rabaud. Hence, the naturalist side of Lapicque should not be underestimated.[77] Even more surprisingly for an institution supposedly devoted to the cultural and sociological approach, Van Gennep took a dim view of all this ecumenical rhetoric, which he thought targeted his own efforts, and found the institute an overly heterogeneous collection still oriented to the "greater glory of craniology."[78]

While the institute was hospitable to Delafosse's articles on religion in western Sudan, to the work of Mauss, to Lévy-Bruhl on numbering among the Australians, and to Rivet on the origins of the name of Peru, it did not sound the death knell for physical anthropology. Verneau read papers on an artificially deformed cranium and discussed pelvis size in a part of France where rickets was less prevalent. Marcel Cohen, a Semitic languages expert, wrote on male breast development in different peoples and dental school lecturer Jules Cloquet discussed dental terminology.[79] Hence for the institute, cultural anthropology had not completely eclipsed the study of significant somatic attributes, even though some of these were clearly not hereditary.

THE NEW SOCIÉTÉ D'ETHNOGRAPHIE

With the patronage of the Colonial Office, the Société d'ethnographie revived in 1913–14 with approximately 100 members. Those involved in colonial lobbying or indigenous affairs in the Ministry of Colonies took an active part. The society did not meet from 1914 until December 1920, when its offices had moved to the headquarters of the Society of Commercial Geography, and, after several prewar issues, the journal

76 Zerilli, "Il Dibattito"; see also Rivet's review of Georges Hervé on miscegenation in Review of Hervé, "Noir et blanc, Le Croisement des races aux Etats-Unis et la théorie de la 'miscegenation,'" 444–5.

77 Zerilli, "Il Dibattito," 247, 252, 254, 262.

78 Van Gennep, "La Signification du 1er Congrès d'ethnographie," 326; also cited in Zerilli, *Il Lato oscuro*, 103.

79 *L'Anthropologie* 23 (1912), supplement.

L'Ethnographie did not resume publication until 1922. A pivotal figure after 1920 was the conservative deputy Louis Marin, who had lectured in ethnography at the private Ecole libre des sciences politiques (founded in 1872 by the philosopher Emile Boutmy) and at another private school featuring the traditionalist doctrines of Frédéric Le Play, the Collège libre des sciences sociales. He was a member of both the old ethnography and anthropology societies, and also taught ethnology to teachers in evening courses at the Ecole d'anthropologie. By 1920 he was administrator of the Collège libre des sciences sociales.

The new Société d'ethnographie would revive an ecumenical definition of ethnography claiming to be as much "sociological as anthropological ... about the life of human groups in the course of evolution." Disavowing Léon de Rosny, even ignoring his role in founding the previous society, they took an opposite approach from van Gennep in including both the "study of human races and the development of civilizations, from the most primitive until the most advanced."[80] The Americanists Hyacinthe de Charencey and Désiré Pectòr, Nicaraguan consuls, were holdovers from the old society, and senator Léon Bourgeois, president of the old society in 1900, received the post of honorary president in the new one. The membership included some from the Trocadéro Muséum of Ethnography staff, the Muséum national d'histoire naturelle director and neo-Lamarckian biologist Edmond Perrier, the conservative and devoutly Catholic Jean Brunhes, professor of human geography at the Collège de France, and Jean-Marie Lanessan, a neo-Lamarckian physician who was a deputy, ex-navy minister, ex-governor of Indochina, and former member of the Société d'anthropologie. Besides Lanessan, the founders included a former minister of colonies, Paul Guieyesse, the senators Lucien Hubert and Henri Bérenger, and the professor of Japanese, Michel Revon, who was a member of the Conseil consultatif des affaires indigènes, which was responsible for protecting indigenous peoples.[81]

Most articles in the journal had cultural content and would represent greater interest in the milieu of peoples. But just as the old ethnographers included physical anthropology despite themselves, so did the new society. They included a note on the structure of the brain in Asiatics by the asylum physician and psychology laboratory director Auguste Marie and another on the "anthropometry of the Malgaches" by Marie and his medical colleague, the internist Léon MacAuliffe.[82] MacAuliffe was content to define a people as a mixture of races living in the same milieu but

80 *L'Ethnographie* 1 (1913): 2.

81 Sibeud, *Une Science impériale*, 109.

82 Marie, "Notes sur la morphologie de l'Encéphale des Asiatiques," 81–8, and Marie and Macauliffe, "L'anthropométrie des Malgaches," 71–2.

was searching for a better definition of race than pigmentation of eyes and hair – the morphology and functioning of respiratory, digestive, muscular, and cerebral systems.

After the war, MacAuliffe and Marie were still studying pigmentation of eyes and hair in French, Italian, and Spanish populations as a means of judicial identification. MacAuliffe would later emerge as a partisan of racial inequality and the importance of heredity for personality.[83] In 1921 Dr Alfred Thooris, chief physician at Villemin hospital in Paris and morphology expert for the War Ministry, also argued that the new key to discerning athletic ability, fitness for military service, or suitability for certain onerous professions was body morphology, including the proportions of trunk to members. He used the term "anthropotechnie," which Manouvrier had proposed to designate practical applications of anthropology. For Thooris, as for nineteenth-century phrenologists, an "Office of Professional Orientation" was needed to guide people to choose an occupation according to this morphology.[84] By 1939 Thooris would be justifying a "moderate racism" and warning against "racial mixing" in the *Revue anthropologique* of the Ecole d'anthropologie.[85]

If the new ethnographers were nationalist, conservative, and leavened with a few hard-line racial theorists, they also embraced an emphasis on the physical, rather than the social, milieu proposed by the human geographer Brunhes. In 1914 Brunhes used race in Le Bon's "historical" sense. "Race is and becomes what history does to it," such as, he thought, when a Serb converted to Islam changes his nationality and therefore his race.[86] Given on a snowy night in June 1914, this lecture was perhaps as dramatic a statement as any in the prewar climate of the potential decline of simplistic old-style racial determinism.

In his 1913 comparison of ethnography and human geography, Brunhes argued that pastoral nomadism, once ascribed to race, was in fact linked to soil and climate, which made transhumance advantageous. Geography therefore "helped determine the distinctive qualities of certain ethnic groups."[87] Ethnography and geography were therefore complementary studies.

Louis Marin did not, however, give up the idea of judging the value of civilizations in his definition of "ethnodicée" in 1922. Ethnography would

83 Paligot, *Races, racisme*, 67.

84 Macauliffe, "Variation de la pigmentation des yeux et des cheveux chez les peuples latins," 31–40; Marie,"Variation de la pigmentation des yeux et des cheveux chez les peuples latins," 41–8; Thooris, "Proportions morphologiques et aptitudes sportives," 25–30.

85 Lebovics, *True France*, 41–2.

86 Brunhes, "Conférence du 8 juin 1914 à la Sorbonne," 15, 17, 22, 24.

87 Brunhes, "Ethnographie et Géographie humaine," 29–40, citation on 31.

be the "science of races," while ethnology would be the "development of civilizations." He formulated a questionnaire intended to describe the "material, intellectual, and social elements of peoples" that would determine the essence of and reasons for the superiority of Western civilization.[88]

From 1923, when Marin became director of the Ecole d'anthropologie, he came to embody the conservative bent of both the new ethnography and anthropology societies. Opposed to the theoretical formulations and socialist politics of the directors of the new Institut d'ethnologie, Marin would be a partisan of the virtues of family life, of rural and regional France, and of a fervent nationalism promoted by his Republican Federation party.[89] In his circumstantial opposition to the socialist Paul Rivet's candidacy to the Muséum anthropology chair in 1928, he would reiterate Rosny's old slogan that the new ethnographers seemed to have previously repudiated, that ethnography "has nothing to do with the sciences of nature," so that a reputed ethnographer like Rivet should not receive a chair in a natural history institution.[90] Marin wished to have his hand in studies of both nature and nurture, but to preserve some distinction. Rather than the old bitterness between ethnographers and anthropologists, the 1920s seemed like an era of potential integration and reconciliation of cultural and physical anthropology.

THE DURKHEIMIANS AND THE INSTITUT D'ETHNOLOGIE

The disdain for the tradition of cranial measurement was most prevalent among the Durkheimians at the Sorbonne. Yet even here there were important nuances to the "discovery of the social." Emile Durkheim himself would not grant all races "an equal native aptitude for all possible forms of mentality."[91] Durkheim thought peoples do have different collective psyches, and was enough of an evolutionist to employ the term "primitive." Still, he insisted no race was condemned to remain at a particular level, though civilization itself created new predispositions unavailable to the less developed. The early Durkheim used organic analogies, even quoting Le Bon in the *Division of Social Labour* as an authority on brain capacity. He also found that the caste system was justifiable by the influence of heredity in "primitive" societies, although it had long since become obsolete. He still believed in hereditary transmission of general abilities even in modern society, but thought these insignificant for

88 Marin, Compte-rendu des séances, 10–11; Lebovics, *True France*, 16, 30–1.

89 Lebovics, *True France*, 37.

90 Zerilli, *Il Lato oscuro*, 152, 154.

91 Durkheim, Review of Boas, "The Mind of the Primitive Man," *Année sociologique* 12 (1909–12): 33, cited in Sibeud, *Une science impériale*, 207–8.

determining specific social roles. But he judged "not all brains are indifferently apt at every function."[92]

On gender issues Durkheim and the team of the *Année sociologique* would never completely disavow a naturalistic interpretation of the status of women. While they professed to see no hierarchies among societies, they clearly thought that more developed societies had more specialization and sexual differentiation in the division of labour. While social rather than biological factors had generated this situation, progressive sexual differentiation was part of a natural order of development.[93] In the contemporary context, they were concerned about women working and becoming so independent as to imperil the stability of marriage and households.

Durkheim and other authors of the periodical therefore raised doubts about the ability of single mothers to raise children, about unlimited no-fault divorce, and opposed free unions without marriage. Durkheim feared that women's participation in the public sphere would require profound changes in the family, "before which one need not recoil, but which one must be able to foresee." Durkheim was willing to allow women into new occupations, particularly in the artistic and literary fields, but was still concerned that juridical equality be confined to women's new functions that relate to an unspecified "women's nature."[94] So even the pre-eminent opponents of an anthropological approach allowed their worries over social instability to manifest concerns about the inherent mental and physical weaknesses of women.

Responding to Marianne Weber's recommendation of freedom for divorce, Durkheim rejected consideration of the family as only a "precarious rapprochement between two beings who at any moment can separate if, and, as long as association lasts, each has their circle of interests and preoccupations." The significance of both the home and marriage would suffer. Even though men suffered more from divorce than women, he still thought conjugal society would be destroyed and many families disorganized.[95] In 1905 at a public discussion he even suggested that the state had an interest in maintaining families too important to leave to the arbitrary will of private parties.[96]

92 Lehmann, "The Question of Caste," 570–1.

93 Charron, "Représentations des rapports sociaux de sexe," 4–5.

94 Durkheim, Review of Jacques Lourbet, "Le Problème des sexes," 364; Charron, "Représentations," 102–3, 123–4, 133–4; Lehmann, "The Question of Caste," 577–8.

95 Durkheim, Review of Marianne Weber, "Ehefrau und Mutter in der Rechtsentwicklung," 363–9, esp. 367–8; see also Durkheim, "Le Divorce par consentement mutuel," 549–54.

96 Durkheim, "Débat sur le mariage et le divorce," 206–15.

As early as the 1901–02 lecture term, a founding member of the Institut d'ethnologie, Marcel Mauss, Durkheim's nephew, in a course on the religions of "uncivilized" peoples at the Ecole pratique des hautes études, would encapsulate a new mood in his contention that "there are no uncivilized peoples, only peoples with a different civilization."[97] Such civilizations were not burdened with the stigma of an unchangeable racial inheritance. The predominance of nurture over nature was a first step toward exorcising the ghost of racial hierarchy.

Yet Mauss's positions on racial hierarchy were not thoroughly consistent. Though Mauss had also joined the Société d'anthropologie in 1905, by 1913 he clearly wished for a reformed ethnology based on social facts, without completely abandoning physical anthropology.[98] In an article in the *Revue de Paris* he called for a Bureau français d'ethnographie or Service d'ethnographie and in a memo addressed to the Public Education Ministry for a Bureau or Institut d'ethnologie (although in the first draft the term was "ethnographie") attached to the University. Mauss claimed that this would aid colonial administrators and institutionalize ethnography.[99] To gain government support and funding, Mauss advocated using ethnography to "manage primitive peoples" in French colonial possessions as well as to study "inferior populations" in the rest of the world.[100] Ethnography would be a "means to prepare the education and march to civilization" of colonized peoples (his proposal excluded European and Eastern peoples). Like van Gennep, he wanted a benevolent domination, a reformed colonialism.[101] Mauss thought ethnography also desirable within the university curriculum to "describe and classify races, peoples, civilizations," especially those of "inferior rank." Description would be a building block for a science of ethnology and useful for study of religion, law, art, and material culture. Clearly he wanted a body intellectually independent of the colonial administration and useful for the "development of humanity."[102]

On the other hand, sociologists wished to reserve the lion's share of theory for themselves with the claim that ethnography was merely useful

97 Mauss, *Œuvres*, 3: 229–30, reprinted from "Leçon d'ouverture à l'enseignement de l'histoire des religions des peuples non-civilisés." *Revue de l'histoire des religions* 45 (1902): 36–55; also cited in Mucchielli, "La dénaturalisation de l'homme," 50.

98 Sibeud, "Projet de présentation d'un bureau d'ethnologie (1913)," 105–15, and Mauss's project for ethnography and ethnology, 117–24.

99 Conklin, "Civil Society," 266–7; Sibeud, *Une Science impériale*, 225; Zerilli, *Il Lato oscuro*, 140, 142.

100 Daniel J. Sherman, "Peoples Ethnographic: Objects, Museums, and the Colonial Inheritance of French Ethnology," 669–703.

101 Sibeud, "Projet," 107, 110–11.

102 Ibid., 117–18, 120.

to circumscribe areas of uniformity and to collect data, while sociologists would furnish the explanatory principles concerning any society. This campaign for a new institute, cut short by the war, resurfaced in 1924. Ultimately, Mauss and the Sorbonne philosopher of religion Lucien Lévy-Bruhl, future president of the council of the institute, approached Paul Rivet, who shared the office of secretary-general with Mauss. With assistance from Delafosse, they would succeed in establishing the Institut d'ethnologie attached to the Sorbonne in 1925. Lévy-Bruhl had ties to Edouard Daladier, minister in the new Cartel des Gauches government, which responded favourably to an appeal from three socialists.[103]

Here Mauss taught descriptive ethnography, Delafosse taught African linguistics and ethnography, Rivet taught physical anthropology and, much later, Claude Lévi-Strauss taught cultural anthropology.[104] There would be an incomplete break with doctrines based on the inherent nature of racial inferiority. Rivet would say that the linguistic and cultural traces of a people were more reliable than physical measurements.[105] Ethnology was no longer the "science of races," as it had been for Broca, but physical anthropology was certainly not abandoned. Lebovics has called this "democratic social science," as compared to Louis Marin's search for the identity of a "true France."[106]

Yet the Institut d'ethnologie, as with Mauss's earlier proposals, had deliberately featured its colonial purpose as well as the "conservation of indigenous civilization by organizations of museums both in France and in the colonies, and ethnological inquiries."[107] Lévy-Bruhl hoped to train colonial administrators, doctors, and missionaries. Students intending to enter colonial service from the Ecole coloniale attended this institute, despite the scientific aspirations of Rivet and Mauss. Now at last administrators were no longer reluctant to demand training in ethnology. Hence, as Conklin argues, the Institut d'ethnologie was implicated in both academic scholarship and the administration of empire.[108]

The opposition between diehard physical anthropologists and the new institute never disappeared, with anatomical anthropologists such as Raoul Anthony even working against Paul Rivet's campaign to succeed

103 Sibeud, *Une Science impériale*, 265–6; Conklin, "The New 'Ethnology' and 'La Situation Coloniale' in Interwar France," 29–46; Sibeud, "Metamorphosis of Ethnology," sees Mauss as a socialist critic of empire aspiring to reforms and discounts the political connections in establishing the institute.

104 See Zerilli, *Il Lato oscuro*, 142 on the atmosphere of the institute; also Laurier, *Paul Rivet*, 341–53.

105 Conklin, "Civil Society," 273.

106 Lebovics, *True France*, 32–6.

107 Lévy-Bruhl cited in Conklin, "The New 'Ethnology,'" 37.

108 Conklin, "The New 'Ethnology,'" 41; and Conklin, "Civil Society," 289–90.

Verneau as Muséum professor of anthropology and Henri Vallois, Rivet's wartime and ultimate successor, minimizing the significance of the institute's work.[109] However, Zerilli has noted a naturalistic heritage from scholars such as Rivet even in the making of the new ethnology. He found that Mauss himself had acknowledged the value of osteology in a "people without history" in a 1925 review. The interaction of race and society was a legitimate subject of study as long as it was not considered an exclusive factor. In Mauss's *Fragment d'un plan de sociologie générale* of 1934, some concession was made to the value of human typology, even though the concept of "race" itself was given only relative value.[110]

Mauss also did not completely give up the idea of a "psychology of peoples." And despite his remarks about there being no such category as really uncivilized peoples, he still believed some people were "low in the ladder of societies." The task of sociology included the relation of social life and human biology, even somatological anthropology. There could be groupings by physiology, blood groups, or musculature, even if not the old cranial osteology. So, as for Marin, ethnology and anthropology were complementary studies, albeit in different proportions.[111] There was no total break in the sense that the sociological paradigm never completely displaced the anthropological perspective, though it was certainly triumphant in the Institut d'ethnologie.

At the 1931 Colonial Exposition Georges Papillault was still differentiating peoples who were assimilable, such as North Africans and Syrians, from Africans and Melanesians who were educable but not assimilable. The stock of the Malgaches of Madagascar and Indochinese Métis could be transformed only by crossing.[112] Hence even reformist colonialism never lost its basis in the idea of racial hierarchy, of indelible natures handicapping all the efforts of nurture.

CONCLUSION

Clearly the Société d'anthropologie soldiered on in the pre-1914 era with hereditarian stalwarts such as Georges Papillault and Georges Hervé, for whom permanence of racial types, racial hierarchy, and the dangers of mixing ("mixophobes," as French historians call them) were major issues. They never gave up on the importance of inherited physical features that correlated with intelligence and character. But perhaps

109 Zerilli, *Il Lato oscuro*, 149–50, 172.

110 Ibid.,173–4, 176–8.

111 Paligot, *La République raciale*, 206, 209–12.

112 l'Estoile, "'Des races non pas inférieures, mais différentes,'" 391–473, esp. 404–5, 423–6.

more representative of the recent trends was the secretary of the society, Manouvrier, for whom the social environment was highly significant.

In the prewar decade already the critique of cranial measurement, the passing of the older generation, and the sensitivity to the excesses of racism after the Dreyfus affair led to a wholesale shift in ethnographic institutions. The re-orientation of outlooks, beginning with Arnold van Gennep, Maurice Delafosse, and Paul Rivet would forever change the tone of ethnography, which from this time would include serious fieldwork. Delafosse and Rivet had earned their anthropological credentials, but cultural and linguistic analysis soon complemented or superseded head measurement.

Delafosse also collaborated with Van Gennep, who represented the drive for autonomy of ethnography from anthropology. In a way the Institut ethnographique international was the revenge of ethnographers for the institutional disarray of Rosny's society and for the anthropologists' superiority complex in the nineteenth century. Nineteenth-century ethnographers and anthropologists were not the polar opposites of the conventional picture, but no doubt anthropologists considered themselves scientific and disdained their literary colleagues. Van Gennep himself was different from hard-line hereditarian anthropologists. He nominally detested denigration of the "primitive" and theories of racial superiority. But even van Gennep did not fully surrender the idea of the arrested development of Algerians. Hence the shift in perspectives was only partial.

The other new institutions all worked in the direction of an ecumenical program reconciling both physical measurements and cultural studies. As a protest against the excessive physicalism of the anthropologists, Rivet and Delafosse helped found the Institut français d'anthropologie, which welcomed the Durkheimian sociologists. The objective was to neglect neither nature nor nurture. Yet the increased emphasis on the cultural and linguistic did not automatically imply a strict egalitarianism. Rivet felt empathy for the plight of Native South Americans forced to endure inequitable social conditions, and he later became a fierce opponent of Nazi-style racism. Yet he never gave up the imperial mission or denied that Europeans had a superior civilization. Similarly, Delafosse saw Africans as capable of progress and creativity, but advocated a reformed colonialism.

The new Société d'ethnographie, backed by the Colonial Office, also aspired to study the cultures of non-Europeans, but welcomed papers in physical anthropology. Some of its members became nationalist conservatives intrigued with racial hierarchy. Partisans of the milieu also heard papers on physical geography, not only on exclusively cultural and social influences.

The most signal triumph of the cultural imperative in ethnology was in the Durkheimians' Institut d'ethnologie, which would lead to the high prestige of French anthropologists in cultural and social studies. But its program did not exclude the inherited attributes of physical anthropology, nor did it surrender its practical mission of being useful to the Colonial Office, which provided a part of its budget. Moreover, even the Durkheimians of the *Année sociologique* had allowed the goal of social stability, though not necessarily hereditarian essentialism, to introduce severe limits to their view of gender roles.

In this institutional change, from 1910 to 1925, the cultural and the social had emerged with new strength but had not completely displaced the penchant for physical measurement of the old anthropology. In the 1930s hard-line physicians and anthropologists advocating racial hierarchy would respond to the new anti-immigrant mood by leavening their theory of race with the importance of culture. However, undesirable cultures, scarcely malleable, could turn out to be victims of the same invidious distinctions as races based on a hierarchy of physical or physiological attributes.

9

Echoes in the Vichy Era

AFTER WORLD WAR I many French intellectuals continued to express concerns about depopulation, low birth rates, and immigration from eastern Europe and from the colonies with non-white populations. Scholars have commented on the continuity of anti-foreign attitudes and racial anti-Semitism during the last years of the Third Republic and their subsequent institutionalization in occupied France and under the Vichy regime.[1] These views were based on the theories of racial and cultural hierarchy that survived in hard-line sectors of anthropology and in the colonial administration. Much of the demographers' or efficiency experts' literature on immigration concerning compatible, assimilable, or productive populations differentiated more desirable European immigrants from undesirable colonial or non-white subjects.[2] Only the extraordinary circumstances of the Vichy era would enable the most extreme proponents of these views to achieve notable influence.

The wider malaise produced renewed concern about the indelible nature of hereditary defects that threatened the survival or purity of the French "race" or ethnic group. The extreme reactions tended to magnify the role of heredity. Yet they were not indifferent to the challenges to racial theory and thus added cultural and psychological identifiers to their new hierarchies. Where no distinguishable biological "race" could be identified the only means to denigrate "undesirable" groups was to attack their culture. As in the nineteenth century and early twentieth centuries, the innate biological dimension could be a more credible

1 Noiriel, *Les Origines Républicaines de Vichy*; Paligot, *Races, racisme et antiracisme*, 127–9; Meyran, "Vichy ou la face cachée de la République," 177–84; on attitudes toward immigration see Schneider, *Quality and Quantity*, 231–92; see Camiscioli, *Reproducing the French Race* on pro-natalism, labour productivity, and racial intermarriage.

2 Camiscioli, *Reproducing the French Race*.

standard for ranking human groups if it complemented or encompassed acquired customs and traditions.

The majority of the French social scientific community would not disavow the trend before 1914 to social and cultural explanations, nor would they be enthusiastic collaborators with the Vichy regime. The anthropologist Paul Rivet would become a vehement anti-racist in the 1930s, even though he still believed in the reality of racial differentiation. A conservative believer in racial and cultural European superiority, Lorrainer Louis Marin, was a staunch enough anti-German nationalist to refuse to bend the Ecole d'anthropologie to the demands of Vichy.[3]

But two notorious physicians and anthropologists converted their scholarly endeavours to the Vichy regime's anti-immigrant and anti-Semitic purposes. The physician and lecturer at both the Ecole d'anthropologie and the Paris Medical Faculty René Martial (1873–1955) began as a social hygienist espousing a largely environmentalist positive eugenicism. By 1933 he used the research of Polish physicians Ludwik and Hanna Hirszfeld and French-based Serbian researcher Nicholas Kossovitch to transform ranking of groups by inherited physical features to ranking by an inherited biochemical index of blood types. Blood types would enable a new selectivity to ensure the supposed successful "grafting" of immigrants onto the host French population. The eventual results were warnings against immigrants from Asia, Africa, or from "Asianized" groups such as many eastern Europeans and most Jews. During the war Martial hoped to revitalize the "French race" by a comprehensive program to regulate marriages and reproduction as well as immigration.[4]

George Montandon (1879–1944), the naturalized Swiss physician and anthropologist who taught at the Ecole d'anthropologie, would develop an anthropological theory of ethnic groups with both cultural and somatic elements permitting him to viciously denounce the Jews.[5] During the late 1930s and during the Vichy regime both Martial and Montandon would popularize the anthroposociology of Vacher de Lapouge.[6]

3 Paligot, *Races*, 49–51.

4 On Martial, see Schneider, *Quality and Quantity*, 231–55, esp. 245–6 on blood types; Schneider, "Hérédité, sang et opposition à l'immigration," 104–17; Taguieff, "Face à l'immigration," 103–131; Taguieff, "La 'Science' du Docteur Martial," 295–332; Camiscioli, *Reproducing the French Race*, 83, 90, 158; Larbiou, "René Martial, 1873–1955," 98–120, and "Le Corps médical," 111–28; Paligot, *Races*, 136–46.

5 Jarnot, "Une Relation récurrente," 17–34; Knobel, "George Montandon et l'ethnoracisme," 277–93.

6 Martial, "Anthropo-biologie des races," 276–303; Martial, "Un Précurseur des groupements sanguins," 620–5; Montandon, "Georges Vacher de Lapouge," 3–7, reprinted in *Georges*[sic] *Montandon et l'Ethnie française.*

RENÉ MARTIAL AND BLOOD GROUPS

Martial first lectured at the Paris Faculty of Medicine in 1929, initially on public health and by 1938 on "anthropobiology," but he did not officially receive a nomination to the new "anthropobiology" chair until 1943. By 1932 he was also an occasional lecturer at the Ecole d'anthropologie."[7] Martial had unmitigated praise for Hippolyte Taine's triad of race, milieu, and moment, and also for Arthur de Gobineau's introduction of racial purity as the most significant indicator of flourishing societies. He also praised Broca for differentiating the viability of offspring of various types of métissage.[8] His quarrel with anthropology was that osteology and somatic types were insufficient to explain the psychology of a race. Blood types, however, remained fixed, while the cephalic index could vary in the same family or in the same people over time.[9] He portrayed himself as a rampart against foolish Lamarckians who would assume that the milieu could easily change a person. The important changes in races occurred through migration and crossing, not adaptation to environment. Martial also, without citing Ribot, used some of the same examples to illustrate the importance of the innate – hereditary diseases and dynasties within professions.[10]

After 1934 Martial would maintain that all races were "resultant races" – none was pure, and each was a compound of biology, history, and psychology. He already used a term employed by Vacher de Lapouge – "ethnie" – later popularized by Deniker, anthropologist Félix Regnault, and George Montandon.[11] A race was an "ensemble of a population whose latent or manifest psychological characteristics (particularly languages) and anthro-biological traits constitute in time (history) a distinct unit."[12] The "ethnie" was a psychological unit more than a common physical type. In fact he considered the French "race" an amalgam of Ligurians, Celts, and Latins (the latter two both Aryan) with less significant subsequent contributions from Franks, Vikings, and Arabs.[13]

7 Schneider, "Hérédité," 106; Paligot, *Races*, 139; Larbiou, "René Martial," 110–12; Taguieff, "La 'Science' du Docteur Martial," 301.

8 Martial, "Politique de l'immigration," 267–94, esp. 284.

9 Martial, *La Race française*, 286; Martial, *Vie et constance des races*, 100.

10 Martial, "Politique de l'immigration," 293; *La Race française*, 296.

11 Martial, *La Race française*, 8; Martial, "Politique de l'immigration," 278; Taguieff, "Immigrés, métis, Juifs," 177–220, esp. 182; on Vacher's use of "ethnie," see Noiriel, *Immigration, antisémitisme et racisme*, 271.

12 Martial, *La Race française*, 317; "Politique de l'immigration," 284; and *Vie et constance des races*, 316.

13 Martial, *La Race française*, 19–32, 66, 94, 109; and *Vie et constance des races*, 123–4.

While claiming not to believe in racial inferiority, he clearly subscribed to the theory of Aryan superiority propounded by Gobineau and Vacher de Lapouge, though not to a German monopoly on Aryanism.[14] Moreover, his remarks on Africans and Jews were not consistent with belief in the equal merits of all groups. He quoted John Ruskin on how education was insignificant given the difference between "well-born and poorly born human creatures, as between a good hunting dog and the vilest bastard cur." He also cited the botanist and anthropologist Reginald Ruggles Gates on how African races were never able to surpass the lower degrees of human culture and offhandedly cited the superior size of brains of Europeans compared to Africans with the implication of superior European intelligence.[15]

In the early 1930s Martial admitted that France needed physically and mentally healthy immigrants both for labour and as a vital catalyst to rejuvenate the French themselves. But even in 1933 he warned that all ethnic minorities were dangerous if not assimilated.[16] Subsequently he based his notion of compatibility largely on the prevalence of certain blood types with Mendelian laws of heredity.[17] Immigration would thus become a science, just as Arthur Bordier had hoped to make colonization a science.[18]

Immigrants with greater prevalence of blood types A or O would have a "biochemical index" closer to the French population, while those with type B, largely from eastern Europe, Asia, or Africa, would undesirably lower the French index should there be intermarriages. He could not yet prove, but seemed convinced, that there was also a weak correlation between the A blood type and the more desirable dolichocephalic head shape, and argued for an even stronger correlation between B types and the less desirable brachycephalic shape.[19] In 1933 anthropologist Georges Papillault was pleased to promote Martial's lectures in the eugenics section of the conservative International Institute of Anthropology.[20]

But Martial was inconsistent in his application of the blood type frontier. Just as there was a Lamarckian prejudice among anthropologists about the malleability of non-Europeans, Martial would have double standards for the biochemical index. In his earlier works, he stressed the

14 Ibid., 75; *La Race française*, 9, 40.

15 Martial, *Vie et constance des races*, 244–5, 251.

16 Martial, "L'Immigration et l'avenir de la France," 257–92, esp. 263, 288.

17 Martial, "Politique de l'immigration," 269–70.

18 Martial, *La Greffe inter-raciale*, cited also in Schneider, *Quality and Quantity*, 244; see summary in Martial, *Vie et constance*, 307–12.

19 Martial, *Vie et constance*, 156, 175; and *Les Métis*, 20.

20 Taguieff, "Face à l'immigration," 116.

importance of racial psychology. Hence, the Poles, despite their lower biochemical index, were suitable immigrants because they had a "Latinized" culture.[21] Similarly, Arabs had a low biochemical index, but represented a great civilization. He judged Berbers a kindred people, and even polygamy in Islamic families had advantages for increasing the population. By contrast, the Armenians had a favourable index but a troublesome Oriental mentality. The Jews were also a "resultant race," not a pure type, and their biochemical index and somatic types varied with their origins. He gave special warnings about their Asiatic or irrevocably mixed nature (except for German or Dutch Jews). But Martial added in an offhand quotation identified merely as coming from an "observer" that the psychology of the Jews was what made them unsuitable. They were unstable complainers, unwilling to remain in the same country or occupation very long, and could not appreciate the beauties of nature.[22] The implication in later works was that the blood type was the foundation of the psychology, since blood was at once physiological and psychological. In 1942 he wrote that however nuanced and subtle psychology was, it was "fixed in races through space and time."[23] By 1939 he could cite a litany of mixed marriages of French people with Jews that produced mentally disturbed or physically challenged offspring due to the unsuitable racial crossing.[24]

In the same year his tone concerning unsuitable immigrants had become more urgent. He warned that African and Asian Métis were unassimilable and lambasted the "myth of equality of races."[25] Unscreened immigration from across the blood type frontier would bring the "physical and psychological disunion of the French race."[26] The conflict of heredities could lead to madness or other symptoms of degeneracy, not the least of which was the "aptitude for discussion, from which springs only confusion, not enlightenment." Taguieff points out the anti-democratic bias in such pronouncements.[27] In any case Martial would send refugees to the French colonies, not to the French homeland.

21 Martial, *La Race française*, 307.

22 Martial, *La Race française*, 216–17 on Armenians, 237n on Jews, 240 on Arabs, 311 on Berbers; Martial, *Vie et constance*, 167; Schneider, *Quality and Quantity*, 245–6; Schneider, "Hérédité," 109–10.

23 Martial, *Vie et constance*, 29; and *Les Métis*, 112–13.

24 Martial, *Vie et constance*, 243–4.

25 Martial, "Etrangers et métis," 517–18; also cited in Camiscioli, "Reproducing the French Race," 122, 136–7.

26 Martial, "Etrangers et métis," 527.

27 Martial, *Vie et constance*, 243, 253–4, 312; Martial, *Les Métis*, 58, 60; Taguieff, "Immigrés," 177–8, 192.

Aside from racial selection, Martial favoured a prenuptial certificate for all marriages within France, to weed out the diseased, long a goal of the Eugenic Society. Most sensationally he advocated sterilization laws for hereditary disease or mental illness, if not as arbitrary as in Germany, at least similar to the model of many American states.[28]

During the Vichy regime Martial wrote popularized reference works and children's books on the "French race" for the Commissariat général des questions juives of Louis Darquier de Pellepoix. By the end of 1942 he became a director of the Institut d'anthroposociologie, where he gave a course named after Vacher de Lapouge's "anthroposociologie" and where Claude, the son of Georges Vacher de Lapouge, was the chairperson. His attempts to lecture at the Medical Faculty on "anthropobiology of races" in 1943 were unsuccessful because of student hostility. But collaborationist notables such as Jacques Doriot would refer to him as a welcome advocate of stricter policies on marriage, sterilization, and protection against racial "pollution" by Jews.[29]

In 1942 his work on *Les Métis* revealed a harsher racism. Not only did he say that unsuitable crossing threaten to make the French a "bastard, ugly, vulgar, degraded race" but also he approved the elder Vacher's idea that the French were less fertile because they were less pure.[30] France was threatened by the "immense Asiatic reservoir" that included the Jews. They were even more dangerous than Asians proper since people trusted them more because they appeared European.[31] Martial would warn ceaselessly about the need to protect the European heritage – never should a French person marry an African, Asian, or Semitic partner and they must fight a defensive war against "Judeo-bolshevism."[32]

To put into effect his selective immigration and eugenic policies, he advocated a massive classification program of non-European foreigners by race according to blood type and cephalic index and of Europeans by nationality. Boards of selection would then send back immigrants of undesirable blood type to their countries of origin, and about 60 per cent of Jews could be sent to their home country or to Palestine. Those that stayed would face prohibition from public office (as Vichy statutes already provided), from voting, and from intermarriage.[33] Unrepentant

28 Martial, *Vie et constance*, 268–9, 279, 282–8.

29 Taguieff, "Face à l'immigration," 124; Taguieff, "La 'Science' du Docteur Martial," 295, 301.

30 Martial, *Les Métis*, 46, 70–1; Yee, "Métissage in France," 415–16, 422.

31 Martial, *Les Métis*, 10, 72; Taguieff, "Immigrés," 207, 210.

32 Martial, *Les Métis*, 180; Taguieff, "Immigrés," 180; Taguieff, "La 'Science' du Docteur Martial," 298, 302.

33 Martial, *Les Métis*, 158–63, 166.

after the war, he would remain an advocate of a fixed psychology of races and continued to lament the instability of racial crossings even in his last book, published in 1955.

Martial was also a vehement opponent of modern feminism. In the introduction and concluding chapter of his Medical Faculty lectures of 1938–39, he ranted against "women who do not take sufficient interest in their role of supporting the family." In orienting women toward politics and careers outside the home, "one has struck the most serious blow at what remained of the instinct of preservation of the family."[34] Motherhood should be the highest honour, according to his philosophy that "life is not a pleasure party, but a long duty." Feminism was a "deformation of the feminine mind" that committed the "biological error" of "disjunction between the organism, commanded by endocrine glands, and the psychology that must result from this physiology," that is, the respect for the feminine functions of guarding the household and raising a family. He hoped men could earn far higher salaries so women could stay home, and that divorce would be made more difficult, though not prohibited.[35] These ideals of the "eternal feminine" prefigured the hierarchical view of the family enshrined during the Vichy regime.[36] In 1942 he continued to rail against the clamorous demands of "Anglo-Saxon feminism exacerbated by Jewish women."[37] His advocacy of household schools for young women harks back to the ideals of Emile Cheysson and the disciples of Frédéric Le Play.[38]

Martial illustrates the persistence of the old anthropological perspective of racial hierarchy now transformed from somatic features or cephalic index to the new scientific idiom of blood types. He never used nature to the exclusion of nurture, since race included "psychology" that could be used to promote anti-Semitism. An undesirable psychology could increase the peril of an incompatible blood type. However, at bottom this was largely a return to nature from culture since the psychology of the biologically distinct and errant group was not considered to be very malleable. Unsuitable blood types would therefore translate into unsuitable races that could not overcome their handicaps. His antagonism to feminism was not based on comparable "scientific" reasoning, but principally on the need to restore a flourishing French population.

34 Martial, *Vie et constance*, 10.
35 Ibid., 293–5.
36 Muel-Dreyfus, *Vichy and the Eternal Feminine*.
37 Martial, *Les Métis*, 92, 219–20.
38 Ibid., 228.

GEORGE MONTANDON AND "ETHNORACISM"

An even more sensational case was the Swiss-born naturalized French physician and anthropologist George Montandon, who from 1919 to 1926 had been an enthusiastic Communist in Switzerland. From 1925 to 1927 he was an assistant and then antagonist to Rivet at the Muséum national d'histoire naturelle in Paris. However, in 1928 Rivet still supported his membership in the Institut français d'anthropologie.[39] Montandon was fully sensitive to the importance of milieu and of the language, religion, and customs of ethnic groups. He eventually relativized the importance of the physical features of "race" so that ethnic groups based on customs and traditions needed equally probing analysis. As an expert on Arctic peoples and the Ainu of southern Sakhalin and northern Japan, he gained enough respect to lecture for Louis Marin in ethnography at the Ecole d'anthropologie in 1931 and in 1933 to succeed Georges Hervé in the ethnology chair.

By 1934 he suggested in time-honoured anthropological fashion that some races could hardly progress, others were "precocious," while only the "late cultures" advanced beyond hunter-gatherers. The "great races" were unequal (aside from Mongoloid, Negroid, and Europoid, he classified Vedd-Australoid and Pygmoid categories). During the war he would argue, like Le Bon, that Europeans unquestionably produced far more superior individuals. He could only speculate about whether psychic qualities corresponded to racial qualities at this point, though he slipped into distinguishing a "Negro mentality" from a white mentality, albeit influenced by cultural milieu. He saw races as purely somatic and hereditary, and measured height, cranial capacity, cephalic index, and prognathism, and observed skin colour, hair colour, eyes, and nose shape as if he were perpetuating the tradition of Paul Broca and Paul Topinard. Like Martial, he mentioned blood groups, but he rejected them as consistent differentiators since they did not match other racial characteristics.[40]

However, like Deniker and Martial, he thought the French had elements of the Blonde, Alp-Armenian, and Brown, or Mediterranean races (similar to Vacher de Lapouge's divisions). During the war these divisions were

39 Knobel, "George Montandon et l'ethno-racisme," 277–9; Paligot, *Races,* 51.

40 Montandon, *La Race, les races,* 13, definition of race; 53–69, characters to measure; 71–2 on blood types; 100 on factors influencing races; Montandon, *L'Ethnie française,* 13, 15 on great races; 20 on differentiation; 36–45, 47 on blood groups and racial characteristic; 148–9 on Negro mentality; Montandon, "L'Aryanisme français," 3–4; and "L'Etudiant français et la science ethnique," 5; see the longer version of Knobel's article, "Antisémitisme et ethno-racisme de George Montandon dans les années 1930 et 1940," at http://www.crif.org/uploads/articles/fichiers/Montandon.pdf; Paligot, *Races, racisme,* 22.

simplified into Nordic, Alpine, and Mediterranean, with the Nordic clearly setting the desirable standard. The French were above all an "ethnie," defined as a "a natural human group, determined by the totality of its characteristics, hereditary (biological) and non-hereditary (traditional)."

For an ethnic group, racial characteristics had to be included and at first had the preponderant role. American "Negroes," he claimed, would otherwise be erroneously classified as belonging to the Anglo-Saxon ethnic group. Indeed he thought ethnology was still defined in France to be the study of races, despite the alleged distortions of Marcel Mauss. But for Montandon linguistic and cultural attributes, including religion, customs, and "mentality," which itself was partly inherited and partly traditional, became increasingly important in the late 1930s and the Vichy era.[41] Only rarely were all three factors – physical features, language, and culture – specific to a particular group.[42]

The Jews were not a distinct race, but rather classified in the Anatolian subrace of the Alp-Armenian, or the Araboid subrace of the Mediterranean or Brown race. Nevertheless, he thought that Jewish "ethnie" did have a recognizable physical type of the "soft parts," not the skeleton, with facial features, particularly nasal shape, overlapping the racial groups. In addition, he thought that social actions helped maintain the type of the Jewish face. This would be a sinister foreshadowing of his later occupation of certifying Jewish identity. In fact, the necessity of adding the importance of psychological, religious, or cultural issues to an anti-Jewish outlook (or to the placing of any phenotype in a hierarchy) was clear, if no definite biological "Jewish race" could be identified. Yet the identification of an inherited ethnic somatic "type" made his wartime activity feasible.[43]

By 1941 he had become a theorist of "ethnoracism," with Aryans at the top of the scale of ethnic groups and Jews at the bottom. Jews were by implication now inassimilable also because of their ethnic, cultural, and religious differences – their behaviour and psychology were as important as physical features. He even suggested that as a tactic for the study of human groups it was better to separate the somatic from the mentality, so as not to have to declare which aspects of mentality were inherited and which were acquired.[44]

41 Montandon, *La Race*, 14–5, 18–20; Montandon, *L'Ethnie française*, 9, 25; Jarnot, "Une Relation récurrente," 18.

42 Montandon, *L'Ethnie française*, 28.

43 Montandon, *La Race*, 262–3; Montandon, *L'Ethnie française*, 139, 144–5; cf. Montandon, "Les Caractères du type judaïque," 15–20.

44 Montandon, "Vacher de Lapouge," 5.

This campaign against the Jews was not just an artifact of the Vichy regime and German occupation, since from 1935 to 1938 Montandon had already metamorphosed into a virulent anti-Semite and opponent of the Popular Front. The reasons are still speculative, but scholars have suggested his desire for popularity in Germany and his search for a marketable theme that might advance his career in the context of cultural xenophobia. By 1938 he was offering bizarre suggestions that Jewish women accused of sexual liaisons with non-Jews should have their noses disfigured so as to be rendered unlikely to reproduce and that Jewish men who had non-Jewish women partners should be castrated.[45] Before the war, he was already advocating in Louis Darquier de Pellepoix's anti-Semitic newspaper the banishment of Jews from France and their confinement in a territory of their own. He continued to promote the expulsion of Jews during the war.[46]

During the Vichy era his collaborator Gérard Mauger vehemently attacked those ethnologists such as Mauss and Rivet who disregarded the "racial fact."[47] Montandon pilloried Jews in 1943 as unproductive intermediaries, asocial exploiters and swindlers of other communities, and noted that their mentality and religion as ethnic identification were more important than physical features for assessing their significance.[48] He, too, worked as an ethnologist for the Commissariat géneral aux questions juives, first for Xavier Vallat, whom he later criticized for not being sufficiently acquainted with the ethno-racist approach to anti-Semitism.[49] He then became head under Darquier de Pellepoix of the Institut d'étude des questions juives et ethnoraciales. Most notably, he would participate in the Aryanization of properties and earn enormous illicit sums of several million francs for conducting nearly 4,000 examinations to certify whether a person was or was not Jewish under the statutes of Vichy. These certificates could lead to or avoid property confiscation, internment, or deportation.[50] Montandon's designation of Jewish religion as merely a sign of ethnic adherence enabled him to identify atheists or other non-practicing Jews as ethnically Jewish under

45 Knobel, "George Montandon," 281–2.

46 Montandon, "L'Aryanisme français," 6; Knobel, "George Montandon," 283.

47 Mauger, "Les Fossoyeurs d'anthropologie," 14–15; Montandon, "L'Etudiant français et la science ethnique," 2.

48 Montandon, "Ethno-raciologie judaïque," 7–8; Jarnot, "Une Relation récurrente," 22–4.

49 Taguieff, "L'antisémitisme à l'époque de Vichy," 45–143, esp. 102.

50 Paligot, *Races*, 56.

Vichy laws.[51] He pushed his reasoning to its logical conclusion as well, since he advocated "extirpation" of the Jewish problem, even including the "death of the flock."[52]

Montandon used the prestige of "scientific" anthropology to trumpet extreme versions of theories of racial and cultural hierarchy that would have been disavowed by most mainstream anthropologists. Of course, anthropologists did not create the anti-Semitic policies of Vichy. The vast majority of republican anthropologists were not routinely anti-Semitic in the nineteenth century. And yet the anthropology of the 1930s was still useful for the re-creation of prestige for the theories of Vacher de Lapouge. In 1943 Vichy literally created the Institut d'anthroposociologie presided by Vacher's son, Claude Vacher de Lapouge. Even with an elastic notion of ethnic group, adding culture to the hereditary factors, Montandon's formulation did not provide protection from the desire to "extirpate" a threat to the "Aryan" French. Belief in some influence of the milieu was insufficient to prevent not only the invidious classification but also a direct incentive to persecution. Resistance members hated Montandon so much that they fatally wounded him in August 1944, and he died in Germany soon after.

Martial and Montandon illustrated the persistence of a firm belief in some circles in the 1930s of racial as well as cultural hierarchy. And while the Durkheimians themselves were not thoroughly emancipated from traces of belief in racial and cultural hierarchy, in the milieu of the Ecole d'anthropologie that belief drifted into more sinister impulses.

51 Montandon, "La Religion juive," 25–6.
52 Montandon, "Ethno-raciologie," 9.

10

Conclusion

THE EMPHASIS PLACED on heredity or milieu is another chapter in the French saga of toleration of "difference." For inherited, indelible differences were less tolerable, and the inability of the milieu to alter substantially people of certain "races" or even European women became for some a philosophical leitmotif. When the milieu allegedly merely induced cultural inferiority, there was, however, no incentive to greater empathy. But we have found no predetermined association between republican universalism or individualism and the exclusion of "difference." Not only were opponents of the Republic inclined to be exclusionary but also republicans themselves had many degrees of reception of the "other."

Even at its most bizarre, the strange mixture of emphasis on heredity and milieu in the Vichy era echoed what had been a familiar pattern since the late nineteenth century. As in modern debates, the almost self-evident interconnection of nature and nurture has shifted the argument about how much each predominates. Answering that question has resulted in turf wars between members of respective disciplinary communities. We have seen how ethnography, anthropology, and sociology at various junctures in the nineteenth and early twentieth centuries all wished to include and absorb the others.

Rosny's ethnographers were students of the language and literature of exotic civilizations who mocked craniology, feared heretical materialism, and adopted an emblem of human brotherhood. Although many were politically conservative and some were even elitist, they proclaimed a set of anti-imperialist principles in 1869. At most other times they cheered on the imperial ventures of France and wished to spread Christianity and European civilization, albeit learning from Asia, Buddhism (for Rosny), and pre-Columbian America. While culture was of primary importance to them, they could not completely rid themselves of stereotypical classifications of non-Europeans. Just below the surface they still accepted physical markers for classifying peoples and believed these characteristics influenced civilizations. Some members openly denigrated Asians and Jews. Indelible nature was not purged from the array of concepts they found palatable. In fact, they wished to encompass anthropology as a subordinate division of ethnography. However, their acceptance of Haitian officers on the African and Oriental Committee suggests they were ready to take some concrete controversial actions to help overcome perceptions of inherited racial handicaps.

The publications and questionnaires of the ethnographers were not well enough organized to make the same impact as the anthropologists' lab, museum, and school: The anthropologists, however republican and politically progressive, were almost always inclined to accept the importance of corporeal differences in a carefully graded hierarchy. They thought the scientific study of bodies and crania led them to unavoidable distinctions. The Muséum school, more open to ethnography and the importance of the milieu, never really surrendered this hierarchical ideal. The development of the most advanced societies inspired many to believe in necessary guidance for the "retarded" who needed to catch up. After some hesitation due to qualms about European adaptability, their confirmation of this hierarchy ultimately elicited endorsement of imperial ventures. Maverick ex-socialist Vacher de Lapouge and conservative Le Bon were extremists in their insistence on the importance of heredity to the practical exclusion of significant environmental influences. In the case of "historical races," Le Bon postulated mostly fixed identities, since non-Europeans would take centuries to improve. Even those like the monogenist Quatrefages, who strongly believed in a single human species, did not question some milder form of racial hierarchy, albeit with potential for change.

For many anthropologists Lamarckianism offered a way out of the hereditarian trap by providing a mechanism for potential improvement – the inheritance of acquired characteristics. However, this mechanism was often vitiated, first by emphasizing the negative acquisitions such as predispositions to physical or mental illness. In addition, the mechanism was not consistently applied. Only Europeans, including European women, could be expected to improve while others were not equally adaptable. Nevertheless, some, like Letourneau, had concluded that non-Europeans were in the end perfectible, albeit not very quickly. The anthropologists were divided on gender issues, with Letourneau ultimately representing the most encouraging for the ideal of equality of women while others insisted that women should adhere to the allegedly biological and social functions of child-rearing and family.

Among the anthropologists a counter-current to the supremacy of heredity was represented by major figures such as Manouvrier and

Louis-Adolphe Bertillon, as well as more obscure adherents such as Lagneau, Schrader, and Durand de Gros. They were much less inclined to denigrate the efforts of ethnographers or to separate sociology from anthropology. For them either the physical environment or, especially for Manouvrier, the social environment, was of crucial importance. Thus, neither born criminals nor born athletes nor inherently inferior races could be countenanced as valid scientific constructs. Vacher de Lapouge's cephalic index was especially unworkable as testimony to superior mind and character. Manouvrier's sociological bent led him to publish this attack in a socialist editor's newspaper. However, Manouvrier never gave up the comparison of brain sizes and would acknowledge European superiority under present conditions. Moreover, his feminism was half-hearted (as was the case of many colleagues on the republican left), as he rejected consideration of women's brain sizes as inferior but believed in a biological necessity that women stay out of competition with men in many professions. Still, the loss of confidence in anthropometry as a discipline with all the answers heralded a critical attitude toward the conventional hierarchical views of race and gender.

The hereditarian anthropological perspective was attractive to some of the leading psychologists in the pioneering years of French empirical psychology. Both Ribot and Binet fell under its spell, in part through the vast influence of Herbert Spencer. Ribot was a moderate republican, anti-clerical by inclination but friendly with conservatives such as Le Bon. For him the anthropological perspective meant attributing maximal significance to inherited dispositions, not only for the negative aspects of illness, but also for Galton's illustrious family dynasties in various professions. For Binet, it meant a fascination with anthropological measurement of heads and other physical features in the hopes of correlating the data with some aspects of intelligence. Both, however, became disillusioned with this approach in the latter part of their careers. For Ribot, the rise of sociology and challenges to the Lamarckianism he supported made him more sensitive to the importance of a collective psychology with social influences. He had even provided the venue for publication of Durkheim's *Rules of Sociological Method*. Ribot at least half-heartedly began to accept the importance of social factors in his study of the psychology and logic of sentiments. For Binet, the lack of significant results led him to wish to develop new methods of measuring intelligence. Though his American interpreters would develop his practices in a direction favouring genetic endowment, he himself would be hesitant to assume that there was a single inherited variable of intelligence.

Whatever Ribot's own views, the periodical he edited, the *Revue philosophique*, manifested similar trends in the philosophical community of a move away from a medical and anthropological fascination with psychological heredity toward a more fitful engagement with sociology. The era of Hippolyte Taine and Herbert Spencer gave way to the era of Durkheim and Célestin Bouglé. The Lamarckianism of some authors was still ambivalent, in that they could be pessimistic in view of inherited dispositions to disease, yet optimistic about the potential for improvement of the species. Philosophers such as Alfred Fouillée and sociologists such as Gaston Richard took the position that the hereditary endowment of race was important in distant ages, while in modern civilization the social milieu was far more significant. Organicist sociologists such as Novicow and Espinas were not necessarily racists, while their opponents Tarde and Bouglé wished to disown a biological approach to sociology. Confirmed Lamarckians like Ribot did not necessarily abandon concepts of racial hierarchy. But others more attracted to Weismann's theory of heredity found just as firm support for a long inherited line of strengths and weaknesses. The backlash against a purely sociological approach, however, was evident in the questions raised by Samuel Jankélévitch and Henri Joly, neither of whom was convinced that the eclipse of heredity by milieu was justified.

The disputes about the social significance of Lamarckianism and Weismann's viewpoint testified to the ideological intrusion into the heredity-milieu question. For those like Le Dantec who wished to combat "aristocratic prejudices," Lamarckianism was a convenient buttress to their arguments because it held out hope for continual improvement by education and transmission of habits. For Bouglé, however, Lamarckian transmissibility created the obstacle of caste, and hence he was willing to see democracy and egalitarianism as social movements opposed to the constraint and hierarchy of Darwinian evolution.

The neglected non-Durkheimian approach to sociology took a more firmly centrist approach than Durkheim's nominally anti-hereditarian stance. We have seen though that on gender issues and thoroughgoing racial equality, even the Durkheimians were more moderate than their reputation. The institutions for the non-Durkheimians established by René Worms were more open to the view that inherited endowment counts for much, but does not determine the trajectory of individuals. However, a hard core of would-be sociologists remained stubbornly committed to hereditarian assumptions despite their primary interests in social facts and issues. On gender issues there was less essentialism about a female nature, but overriding concerns that family and social stability would be threatened if too many freedoms were extended to women. There was no necessary correlation between intransigence on race and intransigence on gender or flexibility on both issues. The early career of the radical feminist Madeleine Pelletier and the anthropological arguments of Clémence Royer illustrated how progressive attitudes on race did not necessarily accompany feminism. However, the Paris Société de sociologie at least admitted women, including radical feminists, to their discussions, and usually gave them a respectful hearing.

In the decade before World War I, a reorganization of societies of ethnography and anthropology lessened the old rivalries among these disciplines and linked them together more closely than ever. The primacy of the cultural and social seemed to triumph in the attention paid to milieu in organizations such as the Institut français d'anthropologie, in van Gennep's various societies and periodicals, and most remarkably in the Institut d'ethnologie at the Sorbonne. However, none of these completely abandoned physical anthropology or completely surrendered notions of racial hierarchy. Meanwhile, mainstream anthropologists continued to discuss the importance of inherited physical characteristics, whether for races or for the descendants of mixed marriages. The new Société d'ethnographie, the Institut français d'anthropologie, and even the Institut d'ethnologie all reconciled studies of the physical and the cultural, though in the latter case the new ethnographic approach created institutional prestige.

So while one of the lessons of the era before World War I was a change from the anthropological perspective in collective psychology, sociology, and a new cultural anthropology, the persistence of hierarchical attitudes into the 1930s is still remarkable. Other scholars have pointed out the consequences for educational manuals and colonial discourse. Hardly any of the social scientists questioned the right of the French to assert domination over the "other" in an empire, nor would they educate those "others" to the same level as French in the mother country. Having an empire is contingent on believing that an empire is necessary. On gender issues, we have also seen only a few exceptions to the general reluctance of most social scientists to grant full equality in the family or in political rights.

The challenge to pure hereditarianism explains the more subtle concepts of race and culture developed in the 1930s. The reconstruction of these concepts shows that the forces of modernity and social change had not overcome traditional hierarchical attitudes. The most extreme heirs of the anthropological perspective in the 1930s, Martial and Montandon, found it necessary to modify the notion of race based on physical characteristics with a linguistic and psychological component that enabled discussion of "ethnic groups." Associating the cultural and social milieu with indelible inheritance served tactically to make their definitions more convincing. But for Martial the new scientific standard of inherited blood types replaced the old osteology with a new ranking unfavourable to Asians, Africans, most Jews, and many eastern Europeans. For Montandon, the psychology of Jews was as detrimental to them as the physical features he claimed to observe. Moreover, the addition of historical and psychological components to race or "ethnie" did not help to protect potential victims, given a belief that they were inassimilable. Vichy notables could use these doctrines as scientific window dressing for racial ideologies adopted for other reasons. Most mainstream French social scientists did not adopt these views, any more than Gobineau and Vacher de Lapouge were honoured names for most anthropologists. But the legacy of scientific ideas that could be put to use by extremists is still sobering. Meanwhile, the gender bias in Vichy continued to see the essential role of women as within the family.

Until the era of decolonization in the 1960s, there was never a consistent and thorough triumph in the French social sciences of nurture over nature and a move away from hierarchies of race and gender. However, at present an egalitarian approach to race and gender is supreme, even if certain marginal political factions have continued to use ideas from the social sciences to promote a hierarchical society.[1] The extent of changed attitudes was indicated in a 1972 law that prohibited discrimination, expressions of hatred, or violence against a person or group based on "membership or non-membership of an ethnic group, nation, race, or religion." Republican universalism had thus travelled far in the postcolonial era, though the laws still refuse to provide statistics on community identities that would make discrimination easier to document.[2] More recently, there has been a flawed but nevertheless moderately successful effort to remove barriers to women holding political office. Since 2000 political parties wishing access to public funds in most electoral situations must nominate women for half the candidacies. Whatever the hesitancies and pitfalls in the adoption of this law, it suggests the eclipse of the partisans of essentialism concerning limits to women's abilities and of exclusive dedication of women to family and household.[3]

But legal texts do not eliminate social and cultural problems. Many of the most heartrending recent political issues fraught with bitterness concern the colonial legacy of France and the integration of citizens of non-European ancestry. The spectre of unquestionable European superiority haunts these issues like an unwelcome presence. Hence the 2005 National

1 Crépon and Mosbah-Natanson, eds. *Les Sciences sociales au prisme de l'extrême droite.*

2 *Journal Officiel de la République française,* 2 juillet 1972, 06803 at http://www.legifrance.gouv.fr/jopdf/common/jo_pdf.jsp?numJO=0&dateJO=19720702&numTexte=&pageDebut=06803&pageFin=.

3 Scott, *Parité.*

Assembly's effort intended to compensate and celebrate Algerians and French soldiers who fought on the French side during the Algerian war led to a proposal to require teaching in school curricula the "positive role of the French overseas presence, notably in north Africa."[4] Over a thousand historians signed a petition opposing a clause in a law that thus "imposes an official lie about crimes, massacres sometimes going as far as genocide, about slavery, about racism inherited from this past."[5] The government effectively removed the offensive clause in 2006.

Yet the former leader of the National Front party, Jean-Marie Le Pen, who garnered 17 per cent of the votes in the second round of the 2002 presidential election, in addition to minimizing the events of the Holocaust in 1987 as a "detail" of World War II, openly proclaimed in 1996 his belief that history demonstrated racial inequality.[6] As recently as 2008, the National Front party program on immigration included a statement that could have been transposed from the pages of René Martial: "at the origin of most of the ills that our country is suffering from, is the policy of immigration conducted by successive governments over the last thirty years."[7] The issues offered for debate there do not openly concern racial hierarchy, but the party has often repeated past appeals to racism. Nominally they are now concerned with questions of employment, social welfare, and assimilation – in short, questions of the compatibility of various cultures and the preservation of a homogeneous French identity.

Beyond the spectrum of French politics, the position of hereditarians today condemns would-be social reformers for violating alleged guidelines for human nature. These "guidelines" are often silent on the means to achieve desirable goals – how to regulate pursuit of self-interest or whether to exert all energies to equalize opportunities. The perennial nature-nurture controversy will not soon disappear because, just as in previous centuries, neither side can claim victory and both sides often have deep-seated political and ideological commitments. In the view of some authors, a theory of dynamic development involving both genetic and environmental factors has displaced the entire nature-nurture opposition that implied separately measurable competing influences.

4 *Journal Officiel de la République française* n°46 du 24 février 2005 page 3128, at http://www.legifrance.gouv.fr/affichTexte.do?cidTexte=JORFTEXT000000448898&dateTexte=

5 "Pétition," http://www.algeria-watch.org/fr/article/hist/colonialisme/loi_23_fevrier_2005.htm

6 Shields, *The Extreme Right in France*, 221; Crépon, "Du Racisme biologique au différentialisme culturel," 187–8n.

7 Front national, Programme, Immigration http://www.frontnational.com/programmeimmigration.php

Moreover, the non-DNA "epigenetic" components in each living cell may affect the expression of genes and be sensitive to environmental influences.[8] In an era when almost all parties to the disputes disavow racism and sexism, however, there is no substitute for trying to separate plausible claims and genuine controversies from exaggerations and fabrications. Understanding allegedly scientific claims in the past about relative influences of heredity and environment is a first step.

8 Shenk, *The Genius in All of Us*, articulately summarizes this viewpoint.

Bibliography

MANUSCRIPT SOURCES

Archives du Collège de France

MS. G-IV-f 45f 1897–1900 Considérations présentées par M. L. Manouvrier à l'appui de sa candidature, 7 nov. 1899

Archives de la Société d'anthropologie de Paris, Musée de l'Homme

C1 2335

1K3 6 mars 1884.

Inventaire, 1L1 and 1K2; correspondance E, letter of July 26, 1888

Registre d'inventaire, 1B1

Registre des procès-verbaux du comité central, 1890–1901, 10 juillet 1890; 12 mars 1891

Archives nationales

62 AJ 5 Assemblée des professeurs, Ecole des langues orientales

AJ 15 556 Dossier d'Ernest-Théodore Hamy

AJ 15 663 Papiers de Hamy

AJ 16 4749 Procès-verbaux de l'Assemblée des professeurs de la Faculté des Lettres, Paris

520 AP 44 Papiers de Binet

F 17 13356 Assemblée des professeurs du Collège de France, 19 janvier 1902, Rapport de Marey

F 17 13558 Ecole d'Anthropologie

F 17 21608 Dossier de Ribot

F 17 23860 Dossier de Léonce-Pierre Manouvrier

Bibliothèque nationale, Manuscrits

N. A. F. 24488, fol. 55, Alfred Binet, Lettre à Louis Havet, 19 août 1899

Muséum national d'histoire naturelle

MS. 2254, Correspondance de Hamy, Lettre de 16 février 1873
MS. 2256,Correspondance de Hamy, Topinard's letter of 4 novembre 1886
MS. 2257, Letter of Verneau to Hamy, 11 septembre 1889 [actually 5 September 1890]
MS. 2258, Letter to Quatrefages of Pruner-Bey, 1 mai 1860, 2 mars 1861, 28 mai 1861
MS. 2312, Letter of Rosny to Hamy 18 juin 1888

PUBLISHED SOURCES

Actes de la Société d'ethnographie 8 (1874–77).
Aldrich, Robert. *Greater France: A History of French Overseas Expansion.* New York: St. Martin's Press, 1996.
Ammon, Otto. "Histoire d'une idée. L'anthropo-sociologie (traduit avec compléments par H. Muffang)." *Revue internationale de sociologie* 6 (1898): 145–71.
Amselle, Jean-Loup and Emmanuelle Sibeud, eds. *Maurice Delafosse, Entre orientalisme et ethnographie.* Paris: Maisonneuve et Larose, 1998.
Andrieu, Bernard. *Alfred Binet, sa vie, son oeuvre.* Paris: Euredit, 2001.
– "Céphalométrie scolaire et intelligence inégale: La Mesure des têtes, un échec à l'origine de l'échelle métrique." In Serge Nicolas and Bernard Andrieu, eds. *La Mesure de l'intelligence (1904–2004),* 157–74. Paris: L'Harmattan, 2005.
Apfelbaum, Erika. "Origines de la psychologie sociale en France." *Revue française de sociologie* 22 (1981): 397–407.
Arréat, Lucien. "Récents travaux sur l'hérédité." *Revue philosophique* 29 (1890): 399–419.
Assézat, Jules. "L'Acclimatement." *Revue d'anthropologie* 4 (1875): 294–305.
Avanzini, Guy. *Alfred Binet et la pédagogie scientifique.* Paris: Vrin, 1974.
Balbi, Adrien. *Introduction à l'Atlas ethnographique.* Paris: Rey and Gravier, 1826.
Barberis, Daniela. "In search of an object: organicist sociology and the reality of society in *fin-de-siècle* France." *History of the Human Sciences,* 16, 3 (2003): 51–72.
Barre, André and Paul Grimanelli. Discussion of Misme, "La Femme de lettres." *Revue internationale de sociologie* 17 (1909): 30–45.
Barthélemy-Madaule, Madeleine. *Lamarck, the Mythical Precursor: A Study of the Relations between Science and Ideology.* Cambridge, Mass.: MIT Press, 1982.
Bauer, Arthur. "Classes sociales." *Revue philosophique* 53 (1902): 275–99.

– "Culture morale et féminisme." *Revue internationale de sociologie* 21 (1913): 288–98.

Beauvois, Eugène de. "Les Skraelings ancêtres des esquimaux." *Revue Orientale et américaine*, new series 3 (1879): 5–48.

Becquemont, Daniel and Laurent Mucchielli. *Le Cas Spencer, Religion, science et politique*. Paris: PUF, 1998.

Bénichou, Claude. "Ribot et L'Hérédité psychologique." In Claude Bénichou, ed. *L'Ordre des caractères: aspects de l'hérédité dans l'histoire des sciences de l'homme*, 73–94. Paris: Vrin, 1989.

Bérenger-Féraud, Laurent-Jean-Baptiste. "Etude sur les Ouolofs (Sénégambie)." *Revue d'anthropologie* 4 (1875): 466–94.

Bernard, Claude. "Exposé des faits et des principes de la physiologie moderne." *Mémoires et travaux de la Société d'ethnographie* 11 (1869): 249–58.

– "Coup d'oeil sur la science ethnographique." *Revue orientale et américaine* 7 (1861–62): 283–7.

Bernard, Félix. Review of August Weismann, "Sélection et hérédité." *Revue philosophique* 34 (1892): 646–56.

Bernard, Noël. "Les populations de la vallée du Mekong." *Revue internationale de sociologie* 15 (1907): 567–89.

Bernardini, Jean-Marc. *Le Darwinisme social en France (1859–1918). Fascination et rejet d'une idéologie*. Paris: CNRS, 1997.

Bertholon, Lucien. "Quel doit être le rôle de la France dans l'Afrique du nord." *Bulletins de la Société d'anthropologie de Paris* 4th series, 8 (1897): 509–36; reply by Zaborowski, discussion by Sanson, 587–604.

Bertillon, Louis-Adolphe. "De l'Influence des milieux ou mésologie." *Bulletins de la Société d'anthropologie de Paris* 2nd series, 7 (1872): 711–28.

– "De l'Influence des milieux sur nos idées et nos moeurs." *Bulletins de la Société d'anthropologie de Paris* 2nd series, 8 (1873): 126–34.

– "De l'Influence du milieu ou mésologie." *La Philosophie positive* 9 (1872): 309–20.

Bertolini, Mara Meletti. *Il Pensiero e la memoria. Filosofia e psicologia nella "Revue philosophique" di Théodule Ribot 1876–1916*. Milano: Franco Angeli, 1991.

Besnard, Philippe. ed. *The Sociological Domain: The Durkheimians and the Founding of French Sociology*. Cambridge: Cambridge University Press, 1983.

Betts, Raymond. *Assimilation and Association in French Colonial Theory 1890–1914*. New York: Columbia University Press, 1961.

Binet, Alfred. "Le Bilan de la psychologie en 1908." *Année psychologique* 15 (1909): v-xii.

– "Ce que vaut l'école primaire, comme préparation à la vie." *Bulletin de la Société libre de l'étude psychologique de l'enfant* 52 (1909): 58–63.

– "La Croissance du crâne et du face chez les normaux entre 4 ans et 18 ans." *Année psychologique* 8 (1902): 345–62.

– *Ecrits psychologiques et pédagogiques*. Guy Avanzini; ed. Paris: Edouard Privat, 1974.
– *L'Etude expérimentale de l'intelligence*. Paris: Schleicher, 1903.
– "Etudes préliminaires de céphalométrie sur 59 enfants d'intelligence inégale choisis dans les écoles primaires de Paris." *Année psychologique* 7 (1901): 369–74.
– "Les Frontières anthropométriques des anormaux." *Bulletin de la Société libre de l'étude psychologique de l'enfant* no. 16 (1904) 430–8.
– *Les Idées modernes sur les enfants*. Paris: Flammarion, 1909.
– *Introduction à la psychologie expérimentale*. Paris: Félix Alcan, 1894.
– "Le Laboratoire de la rue Grange aux Belles; Attitude correcte pour écrire." *Bulletin de la Société libre pour l'étude de la psychologie de l'enfant* no. 34 (1906): 10–24.
– "Lecture d'une physionomie d'enfant." *Bulletin de la Société libre pour l'étude psychologique de l'enfant* no. 32 (1906): 176.
– "La Mesure en psychologie individuelle." *Année psychologique* 4 (1898): 113–23.
– *Modern Ideas about Children*. Suzanne Heisler, trans., Albi: Presses de l'Atelier Graphique Saint-Jean, 1984.
– "Note sur l'étude de Mme. Rousson, La lecture de la physionomie." *Bulletin de la Société libre pour l'étude psychologique de l'enfant* no. 25 (1905) 700–2.
– "Nouvelles recherches de céphalométrie." *Année psychologique* 8 (1902): 341–4.
– "Nouvelles recherches sur la mesure du niveau intellectuel des anormaux." *Année psychologique* 17 (1911): 145–201.
– "A propos de la mesure de l'intelligence."*Année psychologique* 11 (1905): 69–82.
– "Psychologie individuelle. La description d'un objet." *Année psychologique* 3 (1897): 296–332.
– "Recherche sur la technique de la mensuration de la tête vivante." *Année psychologique* 7 (1901) 7: 314–68.
– "Recherches de céphalométrie sur 26 enfants d'élite et enfants arriérés des écoles primaires de Seine et Marne." *Année psychologique* 7 (1901): 403–11.
– "Recherches de céphalométrie sur des enfants d'élite et arriérés des écoles primaires de Paris (déc. 1900 et janvier 1901)." *Année psychologique* 7 (1901): 412–29.
– "Résumé des conférences de Balbiani sur 'Les théories modernes de la reproduction et de l'heredité.'" *Revue philosophique* 26 (1888): 529–59.
– "Les Signes physiques de l'intelligence chez les enfants." *Année psychologique* 16 (1910): 1–30.
Binet, Alfred and Victor Henri. "La psychologie individuelle, Revue génerale." *Année psychologique* 2 (1896): 411–65.

Binet, Alfred and Théodore Simon. "Application des méthodes nouvelles au diagnostic du niveau intellectuel chez les enfants normaux et anormaux d'hospice et d'école primaire." *Année psychologique* 11 (1905): 245–336.

– "Le Développement de l'intelligence chez les enfants." *Année psychologique* 14 (1908): 1–94.

– *Les Enfants anormaux* [1907]. Toulouse: Privat, 1978.

– "Méthodes nouvelles pour le diagnostic du niveau intellectuel des anormaux." *Année psychologique* 11 (1905): 191–244.

Binet, Alfred, Théodore Simon, and V. Vaney. "Le Laboratoire de la rue Grange-aux-belles: Travaux du laboratoire en 1905–06. Première année de sa création." *Bulletin de la Société libre pour l'étude de la psychologie de l'enfant* no. 34 (1906): 10–24.

– "La Misère physiologique et la misère sociale." *Année psychologique* 12 (1906): 1–24.

–"Recherches de pédagogie scientifique." *Année psychologique* 12 (1906): 233–74.

Binet, Alfred and Nicolas Vaschide. "Corrélation des épreuves physiques." *Année psychologique* 4 (1898): 142–72.

– "Corrélation des tests de force physique." *Année psychologique* 4 (1898): 236–44.

– "Historique des recherches sur la céphalométrie."*Année psychologique* 5 (1899): 245–98.

Blanckaert, Claude. "L'Anthropologie au féminin: Clémence Royer (1830–1902)." *Revue de synthèse* 3rd series, no. 105 (janv.-mars 1982): 23–38.

– "L'Anthropologie lamarckienne à la fin du XIXe siècle, Matérialisme scientifique et mésologie sociale," 611–29. In Goulven Laurent, ed. *Jean-Baptiste Lamarck 1744–1829*. Paris: Editions du CTHS, 1997.

– "'Les Bas-fonds de la science française', Clémence Royer. L'Origine de l'homme et le darwinisme social." *Bulletins et mémoires de la Société d'anthropologie de Paris* new series 3, no. 1–2 (1991): 115–30.

– "La Création de la chaire d'anthropologie du Muséum dans son contexte institutionnel et intellectuel 1832–1855," 85–123. In Claude Blanckaert et al., eds. *Le Muséum au premier siècle de son histoire.* Paris: Muséum national d'histoire naturelle, Archives 1997.

– "La Crise de l'Anthropométrie: Des arts anthropotechniques aux dérives militantes (1860–1920)," 95–172. In Claude Blanckaert, ed. *Les Politiques de l'anthropologie.* Paris: L'Harmattan, 2001.

– "Les Deux parties du problème, Renan et l'ethnographie (1848–1865)." *Etudes renaniennes* 102 (1996): 62–89.

– "Of Monstrous Métis? Hybridity, Fear of Miscegenation, and Patriotism from Buffon to Paul Broca," 42–69. In Sue Peabody and Tyler Stovall, eds. *The Color of Liberty: Histories of Race in France.* Durham, N. C.: Duke University Press, 2003.

– "La Naturalisation de l'homme de Linné à Darwin. Archéologie du débat nature/culture," 15–24. In Albert Ducros, Jacqueline Ducros, and Frédéric Joulian, eds. *La culture est-elle naturelle?* Paris: Errance, 1998.
– *La Nature de la société, Organicisme et sciences sociales au XIXe siècle.* Paris: L'Harmattan, 2004.
– "On the Origins of French Ethnology: William Edwards and the Doctrine of Race," 18–55. In George W. Stocking, Jr, ed. *Bones, Bodies, Behavior: Essays on Biological Anthropology.* Madison: University of Wisconsin Press, 1988.
– ed. *Les Politiques de l'anthropologie, Discours et pratiques en France (1860–1940).* Paris: L'Harmattan, 2001.
– "Préface: 'L'anthropologie personnifié,' Paul Broca et la biologie humaine," i-xliii. In Paul Broca, *Mémoires d'anthropologie.* Paris: Jean-Michel Place, 1989.
– "Le Premesse dell'antropologia 'culturale' in Francia. Il dibattito sul 'Questionnaire de Sociologie et d'Ethnographie' di Charles Letourneau (1882–1883)." *La ricerca folklorica* 32 (1995): 51–70.
– *De la Race à l'évolution, Paul Broca et l'anthropologie française. (1850–1900).* Paris: L'Harmattan, 2009.
– "Des Sauvages en pays civilisé, L'anthropologie des criminels (1850–1900)," 55–88. In Laurent Mucchielli, ed. *Histoire de la Criminologie française.* Paris: L'Harmattan, 1994.
– "'Story' et 'History' de l'ethnologie." *Revue de synthèse,* 4th séries, nos 3–4 (1988): 451–67.
– "La Valeur de l'homme: l'idée de nature humaine chez Buffon," 583–600. In J. Gayon, ed. *Buffon 88. Actes du colloque international, Paris-Montbard-Dijon, 1988.* Paris: Vrin, 1992.
Bloch, Adolphe. "L'Intelligence est-elle en rapport avec le volume du cerveau?" *Revue d'anthropologie* 3rd series, 8 (1885): 577–619.
– "De L'Origine des brachycéphales néolithiques de la France." *L'Anthropologie* 12 (1901): 541–7.
– "Sur une Prétendue découverte anthropologique aux Etats-Unis." *Bulletins et mémoires de la Société d'anthropologie de Paris* 6, 2 (1911): 206–7.
– "De la Transformation d'une race dolichocéphale en une race brachycéphale, et vice versa." *Bulletins et mémoires de la Société d'anthropologie de Paris* 5, 2 (1901): 73–83.
Bonneau, Alexandre. "Le Droit de conquête." *Revue orientale et américaine* 3 (1859–60): 371–5.
– *Haiti, ses progrès, son avenir.* Paris: E. Dentu, 1862.
Bordier, Arthur. *La Colonisation scientifique et les colonies françaises.* Paris: Reinwald, 1884.
– "L'Hérédité. Mécanismes et Théories." *Revue mensuelle de l'école d'anthropologie* 4 (1894): 313–28.

– "Le Milieu intérieur et l'acclimatation." *Revue mensuelle de l'école d'anthropologie* 1 (1891): 129–50.

Borlandi, Massimo. "Gabriel Tarde et les criminologues italiens de son temps, à partir de sa correspondance inédite ou retrouvée." *Revue d'histoire des sciences humaines* 3 (2000): 7–56.

Bornstein, Marc H. "Mesure de l'intelligence et pédagogie," 148–72. In Paul Fraisse and Juan Ségui, eds. *Les Origines de la psychologie scientifique.* Paris: PUF, 1994.

Bouglé, Célestin. "Le Procès de la sociologie biologique." *Revue philosophique* 52 (1901): 121–46.

– "La Sociologie biologique et le régime des castes." *Revue philosophique* 49 (1900): 337–52.

Bourgoing, Paul de, ed. *Tableau de la Cochinchine.* Paris: A. Le Chevalier, 1862.

Broca, Paul. "Congrès international des sciences anthropologiques au Palais du Trocadéro … 1878." *Revue d'anthropologie* 2nd series, 1 (1878): 747–53.

– "Discussion sur l'ethnologie et l'ethnographie." *Bulletins de la Société d'anthropologie de Paris* 2nd series, 11 (1876): 216–23.

– *Mémoires d'anthropologie* [1871–88] Claude Blanckaert, ed. Paris: Jean-Michel Place, 1989.

– *On the Phenomena of Hybridity in the Genus Homo.* London: Longman, Green, Longman & Roberts, 1864.

– "Recherches sur l'ethnologie de la France." *Bulletins de la Société d'anthropologie de Paris* 1 (1859): 6–15.

– "Sur le Volume et la forme du cerveau suivant les individus et suivant les races." *Bulletins de la Société d'anthropologie de Paris* 2 (1861): 139–207, 301–21, 441–6.

Brooks III, John I. *The Eclectic Legacy: Academic Philosophy and the Human Sciences in Nineteenth Century France.* Wilmington: University of Delaware Press, 1998.

– "Philosophy and Psychology at the Sorbonne, 1885–1913." *Journal of the History of the Behavioral Sciences* 29 (1993): 123–45.

Brubaker, Rogers. *Citizenship and Nationhood in France and Germany.* Cambridge, Mass.: Harvard University Press, 1992.

Brunel, L[ouis], Ch[arles] Mourey, and [Paul] Chemin-Dupontès, Review of *L'année coloniale. Revue internationale de sociologie* 14 (1906): 169.

Brunhes, Jean. "Conférence du 8 juin 1914 à la Sorbonne." *L'Ethnographie* 1, no. 4 (1914): 1–24.

– "Ethnographie et géographie humaine." *L'Ethnographie* 1 (1913): 29–40.

Bulletins de la Société d'anthropologie de Paris.

Burbridge, David. "Francis Galton on twins, heredity and social class." *British Journal for the History of Science* 34 (2001): 323–40.

Burkhardt, Jr, Richard. *The Spirit of System: Lamarck and Evolutionary Biology.* Cambridge, Mass.: Harvard University Press, 1977.

Caillaud, Gaston. "Les populations du Cambodge." *Revue internationale de sociologie* 15 (1907): 590–6.

Calmette, Emile et al. "Projet de questionnaire concernant les caractères ethniques particuliers du système reproducteur des diverses races." *Actes de la Société d'ethnographie* 7 (1871–4): 13–26.

Camiscioli, Elisa. *Reproducing the French Race: Immigration, Intimacy, and Embodiment in the Early Twentieth Century*. Durham: Duke University Press, 2009.

– "Reproducing the French Race: Immigration, Reproduction, and National Identity in France, 1900–1939." Ph. D. dissertation, University of Chicago, 2000.

Candolle, Alphonse de. *Histoire des sciences et des savants depuis deux siècles d'après l'opinion des principales académies ou sociétés scientifiques*. Genève: H. Georg, 1873.

– *Histoire des sciences et des savants depuis deux siècles d'après l'opinion des principales académies ou sociétés scientifiques*. Bruno Latour, ed. Paris: Fayard, 1987, reprinted from 1885 edition.

– "Lettres sur un questionnaire d'hérédité psychologique." *Revue philosophique* 23 (1887): 659–63.

Carneiro, Robert L. "Herbert Spencer as an Anthropologist." *The Journal of Libertarian Studies* 5, 2 (1981): 153–210.

Caro, Elme. "L'Hérédité psychologique." *Journal des savants* no. 90 (janvier 1874): 50–66.

– "L'Hérédité psychologique, par Th. Ribot." *Séances et travaux de l'Académie des sciences morales et politiques* 101, 1 (1874): 536–40.

Carpintero, Helio. and Javier Molto. "*L'Annèe psychologique* et la naissance de la psychologie expérimentale en France (1890–1945)," 15–34. In Paul Fraisse and Juan Ségui, eds. *Les Origines de la psychologie scientifique, Centième anniversaire de l'Année psychologique, 1894–1994*. Paris: Presses universitaires de France, 1994.

Carroy, Jacqueline and Régine Plas. *Histoire de la psychologie en France XIXe-XXe siècles*. Paris: La Découverte, 2006.

– "La Méthode pathologique et les origines de la psychologie française du XIXe siècle." *Revue internationale de psychopathologie* 12 (1993): 603–12.

– "La Psychologie: Science naturelle et science morale? Lettres inédites de Théodule Ribot à Henri Piéron." *Revue philosophique* 130 (2005): 335–56.

Carson, John. *The Measure of Merit: Talents, Intelligence, and Inequality in the French and American Republics, 1750–1940*. Princeton: Princeton University Press, 2007.

Castaing, Alphonse. "Classification de l'homme dans la Nature." *Revue orientale et américaine* 6 (1861): 167–88, 200–42.

– *Classification de l'homme dans la Nature (Rapport fait à la Société d'Ethnographie)* extrait de la *Revue orientale et américaine*. Paris: Challamel aîné, 1862.

– *Ethnographie de la France*. Paris: Maisonneuve frères et Charles Leclerc, 1885.

– "L'Ethnographie et la structure de la peau humaine," *Revue orientale et américaine* 4 (1860): 119–38.
– "Les Fonctions intellectuelles." *Bulletin de la Société d'ethnographie* 2nd series, 2 (1887): 281–92; 2nd series, 3(1888): 7–22.
– "L'Unité de l'Espèce humaine et les ethnographes des Etats-Unis." *Revue orientale et américaine* 2 (1859): 389–430.
– Review of Xavier Eymar, "Les Peaux-Rouges." *Revue orientale et américaine* 5 (1861): 85–98.
Centlivres, Pierre and Jacques Hainard, eds. *Les Rites de passage aujourd'hui. Actes du colloque de Neuchâtel, 1981*. Lausanne: L'Age d'Homme, 1986.
Centlivres, Pierre and Philippe Vaucher. "Les Tribulations d'un ethnographe en Suisse." *Gradhiva* 15 (1994): 89–101.
Chailleu, Luc. "Histoire de la société d'ethnographie, La Revue orientale et américaine (1858–1879), ethnographie, orientalisme, et américanisme au XIXe siècle." *L'Ethnographie* 86, 1 (1990): 89–107.
– "Léon de Rosny et la connaissance du Japon en France. (Eléments d'une archéologie du 'savoir japonologique' français)." *L'Ethnographie* 86, 2 (1990): 85–107.
Chapuis, Elisabeth. "Binet, une passion de la mesure." *Cahiers Binet-Simon* 650 (1997–1): 71–81.
Charencey, Hyacinthe de. "Le Déluge et les livres bibliques." Extrait de la *Revue orientale et américaine*. Paris: Challamel ainé, 1858 [Review of Schoebel, De l'Universalité du deluge]
– Review of Joseph Henry Dunant, "Notice sur la Régence de Tunis." *Revue orientale et américaine*. 1 (1859): 291.
Charlton, Donald G. *Positivist Thought in France during the Second Empire, 1852–1870*. Oxford: Oxford University Press, 1959.
Charron, Hélène. "Représentations des rapports sociaux de sexe et naturalisme dans le savoir sociologique de la Belle Epoque en France: Une analyse de l'*Année sociologique* (1898–1913)." MA Thesis, Université Laval, 2004.
Chaslin, Philippe. Review of Georges Génil-Perrin, "Histoire des origines et de l'évolution de l'idée de dégénérescence en médecine mentale." *Revue philosophique* 76 (1913): 544–5.
Chauffard, Emile. "Les Populations indigènes du Cambodge et du Laos." *Revue internationale de sociologie* 15 (1907): 561–6.
– Review of Hubertine Auclert, "Le Vote des femmes." *Revue internationale de sociologie* 16 (1908): 718.
– Review of Joran, "Au Coeur du féminisme." *Revue internationale de sociologie* 17 (1909): 218–9.
– Review of Joran, "La Trouée féministe." *Revue internationale de sociologie* 18 (1910): 468–9.

– Review of Pelletier, "L'Emancipation sexuelle de la femme." *Revue internationale de sociologie* 21 (1913): 739–40.
– Review of Pelletier, "La Femme en lutte pour ses droits." *Revue internationale de sociologie* 16 (1908): 235–6.
Cheysson, Emile. "Le Rôle social de la femme" and Discussion. *Revue internationale de sociologie* 13 (1905): 882–98; 14 (1906): 44–65.
Clark, Linda L. *Social Darwinism in France.* Montgomery: University of Alabama Press, 1984.
Clark, Terry N. "Marginality, Eclecticism, and Innovation: René Worms and the *Revue internationale de sociologie* from 1893 to 1914." *Revue internationale de sociologie* 2nd series, 3 (1967): 12–27.
Closson, Carlos. "Dissociation par déplacement." *Revue internationale de sociologie* 4 (1896): 211–22.
Coffin, Jean-Christophe. "Heredity, Milieu and Sin: the Works of Bénédict-Augustin Morel 1809–1873," 153–64. In Hans-Jörg Rheinberger & Staffan Müller-Wille. eds. *The Cultural History of Heredity II: 18th and 19th Centuries.* Berlin: Max Planck Institute, 2003.
– *La Transmission de la folie, 1850–1914.* Paris: L'Harmattan, 2003.
Cohen, Marcel. "Sur l'ethnologie en France." *La Pensée* no. 105 (1962): 85–96.
Compayré, Gabriel. "Les Principes de l'éducation d'après Herbert Spencer." *Revue philosophique* 3 (1877): 168–85.
Comptes rendus des séances de la Société d'ethnographie.
Congrès international des sciences ethnographiques 15–17 juillet 1878. Paris: Imprimerie nationale, 1881.
Congrès international des sciences ethnographiques ... 1889. Paris: Imprimerie nationale, 1890.
Congrès international des sciences ethnographiques, 1889, Procès-verbaux sommaires. Paris: Imprimerie nationale, 1890.
Conklin, Alice. "Civil Society, Science, and Empire in Late Republican France: The Foundation of Paris's Museum of Man." *Osiris* 17 (2002): 255–90.
– "Colonialism and Human Rights, A Contradiction in Terms? The Case of France and West Africa, 1895–1914." *American Historical Review* 103, 2 (1998): 419–42.
– *A Mission to Civilize: The Republican Idea of Empire in France and West Africa 1895–1930.* Stanford: Stanford University Press, 1997.
– "The New 'Ethnology' and 'La Situation Coloniale' in Interwar France." *French Politics, Culture, and Society* 20 (2002): 29–46.
Corre, Armand. "De l'Acclimatement dans la race noire africaine." *Revue d'anthropologie* 2nd series, 5 (1882): 31–97.
Cosentini, Francesco. "Féminisme et science positive." *Revue internationale de sociologie* 17 (1909): 711–35, 809–36.

Coste, Adolphe. *Les Conditions du bonheur et de la force pour les peuples et les individus.* Paris: Guillaumin, 1879.

– Discussion of Golberg, "L'Origine des races et la division du travail." *Annales de l'Institut international de sociologie 1895* 2 (1896): 369.

– Discussion of Roberty, "Les Préjugés de la sociologie contemporaine." *Annales de l'institut international de sociologie* 1900 7 (1901): 270–2.

Cowan, Ruth Schwartz. *Sir Francis Galton and the Study of Heredity in the Nineteenth Century.* New York: Garland, 1985.

– "Nature and Nurture: The Interplay of Biology and Politics in the Work of Francis Galton." *Studies in the History of Biology* 1 (1977): 133–208.

Crapez, Marc. *La Gauche réactionnaire, Mythes de la plèbe et de la race.* Paris: Berg, 1998.

Crépon, Sylvain. "Du Racisme biologique au différentialisme culturel: les sources anthropologiques du GRECEL." 151–90. In Sylvain Crépon and Sébastien Mosbah-Natanson, *Les Sciences socials au prisme de l'extrême droite.* Paris: L'Harmattan, 2008.

Crépon, Sylvain and Sébastien Mosbah-Natanson, eds. *Les Sciences sociales au prisme de l'extrême droite.* Paris: L'Harmattan, 2008.

Dally, Eugène. "Femmes." *Dictionnaire encyclopédique des sciences médicales,* 4th series, F-K 1, 427–39. Paris: G. Masson and P. Asselin, 1877.

– "Histoire d'ethnologie." Summary by Ludovic Martinet. *Revue géographique internationale* 2 (1877): 27–9, 61–2.

Daniel, Gaston. "Essai anthropologique sur les nègres et les mulâtres du Congo." *Revue anthropologique* 24 (1914): 356–71.

Delafosse, Maurice. "Les Agni." *L'Anthropologie* 4 (1893): 402–24.

– "Les Libériens et les Baoulés, Nègres dit civilisés et nègres dits sauvages." *Les milieux et les races* 1 (1901): 97–122, 129–40.

– *Les Noirs de l'Afrique.* Paris: Payot, 1922.

– Review of Jules Harmand, "Domination et colonisation." *Revue ethnographique et sociologique* 5 (1912): 65–6.

– "Les Vaï, Leur langue et leur système d'écriture." *L'Anthropologie* 10 (1899): 129–51.

Delage, Yves. "La Nouvelle théorie de l'hérédité de M. Weismann." *Revue philosophique* 35 (1893): 561–89.

Delbet, Ernest. Discussion of Limousin, "Question monétaire à propos de l'Exposition de 1900." *Revue internationale de sociologie* 9 (1901): 289.

– Discussion of Ponsinet and La Grasserie, "Le Rôle social de la femme." *Revue internationale de sociologie* 14 (1906): 134–41.

Delisle, Fernand. Review of Emile Houzé, "L'Aryen et l'anthroposociologie; étude critique." *L'Anthropologie* 17 (1906): 698–700.

Delvaille, Jules. Review of Richard, "La Femme dans l'histoire." *Revue philosophique* 70 (1910): 412–17.

– Review of Worms, "Philosophie des sciences sociales, t. 1." *Revue philosophique* 56 (1903): 661–6.
– Review of Worms, "Philosophie des sciences sociales, t. 2." *Revue philosophique* 59 (1905): 529–31.
– Review of Worms, "Philosophie des sciences sociales, t. 3." *Revue philosophique* 63 (1907): 651–5.
– "La Vie sociale." *Revue philosophique* 58 (1904): 583–601.
Deniker, Joseph. "La Question des races en psychologie." *L'Année psychologique* 13 (1907): 292–307.
– "Les Races et les peuples d'Europe." *Revue mensuelle de l'école d'anthropologie* 8 (1898): 198–206.
– *Les Races et les peuples de la terre.* Paris: C. Reinwald, 1900.
– *Les Races et les peuples de la terre.* Paris: Masson, 1926.
– Review of Verneau, "Les Races humaines," *L'Anthropologie* 2 (1891): 386.
Desdouits, Théodore. "Le Progrès moral et ses conditions." *Revue orientale et américaine* 2nd series, 6 (1892): 146–63.
Desmet, Piet. "Abel Hovelacque et l'école de linguistique naturaliste: De l'inégalité des langues à l'inégalité des races," 55–93. In Claude Blanckaert, ed. *Les Politiques d'anthropologie.* Paris: L'Harmattan, 2001.
Diamond, Jared. *Guns, Germs, and Steel: The Fates of Human Societies* [1997]. New York: Norton, 1999.
Dias, Nélia. *La Mesure des sens: Les anthropologues et le corps humain au XIXe siècle.* Paris: Aubier, 2004.
– *Le Musée d'ethnographie du Trocadéro (1878–1908).* Paris: Editions du CNRS 1991.
– "Une Science nouvelle: la géo-ethnographie de Jomard," 159–83. In Marie-Noëlle Bourguet, Bernard Lepetit, Daniel Nordman, Maroula Sinarellis, *Invention scientifique de la Méditerranée. Egypte, Morée, Algérie.* Paris: Editions de l'EHESS, 1998.
DiCristina, Bruce. "Durkheim's Latent Theory of Gender and Homicide." *British Journal of Criminology* 46 (2006): 212–33.
Diriwächter, Rainer. "Völkerpsychologie: The Synthesis that never was." *Culture and Psychology* 10 (2004): 85–109.
Dissard, Clotilde. "L'Avenir du féminisme." *Revue internationale de sociologie* 5 (1897): 937.
– "Le Congrès féministe de Paris en 1896." *Revue internationale de sociologie* 4 (1896): 537.
Dorlin, Elsa. *La Matrice de la race, généalogie sexuelle et coloniale de la nation française.* Paris: La Découverte, 2006.
Douarche, Léon. "La Vie sociale aux Etats-Unis." *Revue internationale de sociologie* 13 (1905): 193–4.
Dufour, Véronique. "Binet et la pédagogie," 163–72. In Paul Fraisse et Juan Ségui, eds. *Les Origines de la psychologie scientifique.* Paris: PUF, 1994.

Dugas, Ludovic. "Une Hérédité psychologique par contraste." *Revue philosophique* 76 (1913): 275–84.
– Review of Fouillée, "Psychologie du peuple français." *Revue philosophique* 46 (1898): 323–6.
Duprat, Guillaume-Léonce. "La Psycho-sociologie." *Revue internationale de sociologie* 22 (1914): 656–77.
– Review of Dagan, "Enquête sur l'antisémitisme." *Revue internationale de sociologie* 7 (1899): 909–10.
– Review of Finot, "Le Préjugé des races." *Revue internationale de sociologie* 13 (1905): 907–8.
– Review of Ludwig Gumplowicz, "La Lutte des races." *Revue internationale de sociologie* 17 (1909): 542–4.
– Review of Paul Lapie, "La Femme dans la Famille." *Revue philosohique* 68 (1908): 663–6.
– Review of Le Bon, "La psychologie politique et la défense sociale." *Revue philosophique* 71 (1911): 325–7.
– Review of Richard, "La Sociologie générale et les lois sociologiques." *Revue philosophique* 74 (1912): 512–6.
Durand de Gros, Joseph. "Sur l'Action des milieux géologiques dans l'Aveyron." *Bulletins de la Société d'anthropologie de Paris* 2nd series, 3 (1868): 135–47.
– "Sur l'Action modificatrice des milieux sur l'homme et sur les animaux." *Bulletins de la Société d'anthropologie de Paris* 2nd series, 3 (1868): 228–40; 242–59.
– *Questions de philosophie morale et sociale.* Paris: Félix Alcan, 1901.
Durkheim, Emile. "Criminalité et santé sociale." *Revue philosophique* 39 (1895): 518–23.
– "Débat sur le mariage et le divorce," *Libres entretiens* de l'Union pour la vérité 2: 206–15. In Durkheim, *Textes.* Paris: Editions du minuit, 1975.
– "Le Divorce par consentement mutuel." *Revue bleue* 44 (1906): 549–54.
– *De la Division du travail social* [1893]. Paris: PUF, 1930, reprint 1991.
– "Les Règles de la méthode sociologique." *Revue philosophique* 37 (1894): 465–98, 577–607; 38 (1894): 14–39, 168–82.
– Review of Jacques Lourbet, "Le Problème des sexes." *L'Année sociologique* 4, 1899–1900 (1901): 364.
– Review of Ribot, "La Logique des sentiments." *L'Année sociologique* 9 (1904–5): 156–8.
– Review of Marianne Weber, "Ehefrau und Mutter in der Rechtsentwicklung." *L'Année sociologique* 11, 1906–9 (1910): 363–9.
– *Le Suicide, Etude de sociologie* [1897]. Paris: Presses universitaires de France, 1990.
Dussaud, René. Review of Van Gennep, "Mythes et légendes de l'Australie." *Revue de l'Ecole d'anthropologie* 20 (1910): 222–3.

– Review of Van Gennep, "Rites de passage." *Revue de l'Ecole d'anthropologie* 19 (1909): 141–2.

Edison, Paul. '"Red-Skins' and White Legends: The Société des Américanistes in the Early Twentieth Century." Annual Meeting of the Society for French Historical Studies, Milwaukee, April, 2003.

– "The Society of Ethnography in the Nineteenth Century: From Moral Brotherhood to Racial Science." Annual Meeting of the Society for French Historical Studies, Toronto, 2002.

Edwards, William Frédéric. *Des Caractères physiologiques des races humaines.* Paris: Compère jeune, 1829.

– "Esquisse de l'état actuel de l'anthropologie." *Mémoires de la Société ethnologique de Paris* 1 (1841): 109–28.

Espinas, Alfred. "Etre ou ne pas être, ou du Postulat de la Sociologie." *Revue philosophique* 51 (1901): 449–80.

Fabiani, Jean-Louis. *Les philosophes de la République.* Paris: Editions du minuit, 1988.

Fabre, Daniel. "Le *Manuel de folklore français* d'Arnold van Gennep," 641–75. In Pierre Nora, ed. *Les lieux de mémoire.* III, *Les Frances,* 2. Paris: Gallimard, 1992.

Fancher, Raymond E. "Alphonse de Candolle, Francis Galton, and the Early History of the Nature-Nurture Controversy." *Journal of the History of the Behavioral Sciences* 19 (1983): 341–52.

– *The Intelligence Men: Makers of the IQ Controversy.* New York: Norton, 1985.

– "Science, Ideology, and the Media: The Cyril Burt Scandal." *Science* 253 (1991): 1565–6.

Fee, Elizabeth. "Nineteenth-Century Craniology: The Study of the Female Skull." *Bulletin of the History of Medicine* 53 (1979): 415–33.

Fogarty, Richard and Michael Osborne. "Constructions and Functions of Race in French Military Medicine, 1830–1920," 206–36. In Sue Peabody and Tyler Stovall, *The Color of Liberty: Histories of Race in France.* Durham: Duke University Press, 2003.

Forrest, Derek William. *Francis Galton, The Life and Work of a Victorian Genius.* London: Paul Elek, 1974.

Foschi, Renato and Elisabetta Ciccioli. "Politics and Naturalism in the 20th Century Psychology of Alfred Binet." *History of Psychology* 9, 4 (2006): 267–89.

Fouillée, Alfred. "Les Facteurs des caractères nationaux." *Revue philosophique* 45 (1898): 1–29.

– "Les Origines de notre structure intellectuelle et cérébrale." *Revue philosophique* 32 (1891): 571–602.

Freeden, Michael. "Eugenics and Progressive Thought: A Study in Ideological Affinity." *The Historical Journal* 22, 3 (1979): 645–71.

Front national, Programme, Immigration http://www.frontnational.com/programmeimmigration.php

Gagey-Macadoo, Mme. Julie. Review of W. E. B. Dubois, "Efforts for Social Betterment among Negro-Americans." *Revue internationale de sociologie* 18 (1910): 699–701.

Galton, Francis. *English Men of Science, Their Nature and Nurture.* Ruth Cowan, ed. London: Macmillan, 1874, reprint London: Frank Cass, 1970.

– "Hereditary Character and Talent." *MacMillan's Magazine* 11 (1864): 157–66 and (1865): 318–27.

– *Hereditary Genius: An Inquiry into Its Laws and Consequences* [1869] London: Macmillan, 1914 (reprint of 2nd ed., 1892).

– "The history of twins, as a criterion of the relative powers of nature and nurture." *Fraser's Magazine* 12 (1875): 566–76, revised in *Journal of the Anthropological Institute* 5 (1876): 391–406.

– Review of Ribot, "Psychological Heredity." *The Academy: A Weekly Review of Literature, Science, and Letters* 7, no. 143, new series (30 January 1875): 118–19.

– "Short notes on heredity." *Journal of the Anthropological Institute* 5 (1875): 324–9.

Garanderie, Antoine de la. "Préface." In Alfred Binet, *Oeuvres complètes, t. 20, L'Etude expérimentale de l'intelligence.* Paris: Euredit, 2001.

Gasser, Jacques. "La Notion de mémoire organique dans l'oeuvre de T. Ribot." *History and Philosophy of the Life Sciences* 10 (1988): 293–313.

Geiger, Roger L. "René Worms, l'organicisme et l'organisation de la sociologie." *Revue française de sociologie* 22 (1981): 345–60.

Georges Montandon et l'Ethnie française. Paris: Archives de l'histoire de l'ethnologie, 1993 (reprint of *L'Ethnie française,* 1941–44).

Gillham, Nicholas W. *A Life of Sir Francis Galton: From African Exploration to the Birth of Eugenics.* New York: Oxford University Press, 2001.

Gissis, Snait. "Herbert Spencer's two editions of the Principles of Psychology: 1855 and 1870/72. Biological heredity and cultural inheritance," 137–52. In Staffan Müller-Wille and Hans-Jörg Rheinberger, eds. *A Cultural History of Heredity III: 19th and Early 20th Centuries.* Berlin: Max Planck Institute for the History of Science, 2005.

– "Late Nineteenth Century Lamarckism and French Sociology." *Perspectives on Science,* 10, 1 (2002): 69–122.

Gley, Eugène. Review of Joseph-Jules Dejerine, "L'Hérédité dans les maladies du système nerveux." *Revue philosophique* 23 (1887): 311–16.

Golberg, Mécislas. "L'Origine des races et la division du travail." *Annales de l'Institut international de sociologie 1895* 2 (1896): 345–85.

Goldstein, Jan. *The Post-Revolutionary Self: Politics and Psyche in France, 1750–1850.* Cambridge, Mass.: Harvard University Press, 2005.

Gordon, Felicia. "Convergence and Conflict: Anthropology, Psychiatry, and Feminism in the Early Writings of Madeleine Pelletier (1874–1939)." *History of Psychiatry* 19 (2008): 141–63.
– *The Integral Feminist, Madeleine Pelletier, 1874–1939*. Minneapolis: University of Minnesota Press, 1990.
Gould, Stephen Jay. "Evolutionary Psychology: An Exchange." *New York Review of Books*, 44, 15 (9 October 1997): 55–6.
– *The Mismeasure of Man*. New York: Norton, 1981.
– *The Mismeasure of Man*. 2nd ed. New York: Norton, 1996.
Grandmaison, Olivier Le Cour. *La République impériale, Politique et racisme d'Etat*. Paris: Fayard, 2009.
Grasserie, Raoul de La. "Principes sociologiques des nationalités." *Revue internationale de sociologie* 13 (1905): 417–66.
– Discussion of Roberty, "Les Préjugés de la sociologie contemporaine." *Annales de l'Institut international de sociologie 1900*. 7 (1901): 267–8.
– "La Race et le milieu comme facteurs de l'évolution sociale." *Revue internationale de sociologie* 10 (1902): 356–94.
– Review of Frederick Hoffman, "Des Traits et des tendances de la race nègre en Amérique." *Revue internationale de sociologie* 6 (1898): 137.
– "Le Rôle de la femme dans le progrès de la civilisation." *Revue internationale de sociologie* 14 (1906): 127–54.
Gravlee, Clarence C. H., Russell Bernard, and William R. Leonard. "Heredity, Environment, and Cranial Form: A Reanalysis of Boas's Immigrant Data." *American Anthropologist* 105, 1 (2003): 125–38; 105, 2 (2003): 326–32.
Grimanelli, Paul. *La Femme et le positivisme*. Paris: Pelletan, 1905.
Guyau, Jean-Marie. "L'Hérédité morale de la théorie de Herbert Spencer." *Revue philosophique* 7 (1879): 308–15.
Guyot, Yves. *Lettres sur la politique coloniale*. Paris: C. Reinwald, 1885.
Hadjian, Joseph. "Emile Durkheim et la taille des crânes." *DEES* 115 (1999): 19–30.
Hammond, Michael. "Anthropology as a Weapon of Social Combat in Late Nineteenth-Century France." *Journal of the History of the Behavioral Sciences* 16 (1980): 118–32.
Hamy, Ernest-Théodore. "De l'Origine et des progrès récents de l'Ethnologie." *Bulletin hebdomadaire de l'Association scientifique de France* 2nd series, no. 212 (20 avril 1884): 37–47.
– "Revue des journaux." *Revue d'anthropologie* 2 (1873): 137–44.
– "Tombouctou." *Revue générale des sciences pures et appliqués* 13 (1902): 957–72.
Harvey, Joy Dorothy. "*Almost a Man of Genius*": *Clémence Royer, Feminism, and Nineteenth-Century Science*. New Brunswick, N.J.: Rutgers University Press, 1997.

– "Races Specified, Evolution Transformed: The Social Context of Scientific Debates Originating in the Société d'anthropologie de Paris, 1859–1902." PhD dissertation, Harvard University, 1983.

Hause, Steven C. "More Minerva than Mars: The French Women's Rights Campaign and the First World War," 97–113. In M. R. Higonnet et al. eds. *Behind the Lines: Gender and the Two World Wars.* New Haven: Yale University Press, 1987.

Hause, Steven and Anne Kenney. *Women's Suffrage and Social Politics in the French Third Republic.* Princeton: Princeton University Press, 1984.

Hecht, Jennifer Michael. *The End of the Soul: Scientific Modernity, Atheism, and Anthropology in France, 1876–1936.* New York: Columbia University Press, 2003.

– "The Solvency of Metaphysics: The Debate over Racial Science and Moral Philosophy in France, 1890–1919." *Isis* 90, 1 (1999): 1–24.

– "Vacher de Lapouge and the Rise of Nazi Science." *Journal of the History of Ideas* 61, 2 (2000): 285–304.

– "A Vigilant Anthropology: Léonce Manouvrier and the Disappearing Numbers." *Journal of the History of the Behavioral Sciences* 33, 3 (1997): 221–40.

Heilbron, Johan. "Sociology and positivism in 19th-century France: the vicissitudes of the Société de Sociologie (1872–4)." *History of the Human Sciences* 22, 4 (2009): 30–62.

Hervé, Georges. "L'Alsace." *Revue de l'Ecole d'Anthropologie* 13 (1903): 285–300; 14 (1904): 296–7.

– "Le Croisement des races aux Etats-Unis et la théorie de la 'miscégénation.'" *Revue de l'Ecole d'anthropologie* 16 (1906): 337–58.

– "Ethnologie," 68–78. In Henri Thulié, ed. *1876–1906 L'Ecole d'anthropologie de Paris.* Paris: Félix Alcan, 1907.

L'Homme, Journal illustré des sciences anthropologiques. Gabriel Mortillet, ed. 4 vols. 1884–87.

Hovelacque, Abel. *Langues, races, nationalités.* Paris: Ernest Leroux, 1873.

– *Langues, races, nationalités.* 2nd ed. Paris: Ernest Leroux, 1875.

Hovelacque, Abel and Georges Hervé. *Précis d'anthropologie.* Paris: Adrien Delahaye and Ernest Lecrosnier, 1887.

Hoyt, David L. "The Reanimation of the Primitive: *Fin-de-Siècle* Ethnographic Discourse in Western Europe." *History of Science* 39 (2001): 331–54.

– "Totemism, Ethnography, and the Crisis of Liberal Politics in the Fin-de-siècle and Belle Epoque." Annual Meeting of Society of French Historical Studies, April 2003, Milwaukee.

Huteau, Michel. "La Réception de l'Echelle Métrique de l'Intelligence en France," 41–72. In Serge Nicolas et Bernard Andrieu, *La Mesure de l'intelligence.* Paris: L'Harmattan, 2005.

Institut français d'anthropologie. Compte-rendu des séances supplément à *L'Anthropologie* 1 (1911).

Issaurat, Cyprien. Review of Ernest Renan, "Qu'est-ce qu'une nation?" *Revue d'anthropologie* 2nd series, 5 (1882): 355–7.
– Review of Clémence Royer, "Origine de l'homme et des sociétés." *Revue d'anthropologie* 4 (1875): 330–41.
– "Sur les Mémoires de la Société d'ethnographie contenus dans la deuxième partie du tome XII." *Bulletins de la Société d'anthropologie de Paris* 2nd series, 10 (1875): 186–92.
Jacoby, Paul. *Etudes sur la sélection dans ses rapports avec l'hérédité chez l'homme.* Paris: Germer Baillière, 1881.
Jankélévitch, Samuel. "Essai de critique sociologique du Darwinisme." *Revue philosophique* 73 (1912): 476–92.
– "La Position actuelle du problème de l'hérédité." *Revue philosophique* 75 (1913): 545–67.
– Review of Bouglé, "Essai sur le régime des castes." *Revue philosophique* 66 (1908): 424–7.
– Review of Novicow, "La Critique du Darwinisme social." *Revue philosophique* 69 (1910): 537–41.
– "Du Rôle des idées dans l'évolution des sociétés." *Revue philosophique* 66 (1908): 256–80.
Jarnot, Sébastien. "Une Relation récurrente: science et racisme, L'exemple de l'*Ethnie française*." *Les Cahiers du Cériem* 5 (2000) 17–34.
Joly, Henri. Review of Jean Finot, "Le Préjugé des races." *Revue philosophique* 61 (1906): 100–7.
Jomard, Edme-François. "Classification méthodique des produits de l'industrie extra-européenne." *Revue orientale et américaine* 7 (1862): 131–7, 241–53.
– *Considérations sur l'objet et les avantages d'une collection spéciale consacrée aux cartes géographiques et aux diverses branches de la géographie*, Paris: E. Duverger, 1831.
Joran, Théodore. "Le Féminisme à l'heure actuelle." *Revue internationale de sociologie* 15 (1907): 321–36.
Journal Officiel de la République française no 46 du 24 février 2005: 3128, http://www.legifrance.gouv.fr/affichTexte.do?cidTexte=JORFTEXT000000444898&dateTexte=
J.-P., Review of Marcel Mauss and Henri Beuchat, "Essai sur les variations saisonnières des sociétés Eskimos." *L'Anthropologie* 17 (1906): 711–14.
– Review of Mauss, "Essai d'une théorie générale de la magie." *L'Anthropologie* 19 (1908): 360–2.
Kahn, Sholom Jacob. *Science and Aesthetic Judgment: a Study in Taine's Critical Method.* New York: Columbia University Press, 1953.
Kalmar, Ivan. "The Völkerpsychologie of Lazarus and Steinthal and the Modern Concept of Culture." *Journal of the History of Ideas* 48, 4 (1987): 671–90.
Karady, Victor. "Durkheim et les débuts de l'ethnologie universitaire." *Actes de la recherche en sciences sociales* 74 (1988): 24–32.

Kellès-Krauz, Kazimierz (Casimir). Discussion of Worms, "La Distinction entre l'art et la science." *Annales de l'Institut international de sociologie 1894* 1 (1895): 179–80.

Ki-Toung, Tcheng. "La Nationalité chinoise, apercu ethnographique," 7–13. In *Congrès international des sciences ethnographiques, Mémoires*. Paris: Ernest Leroux, 1892.

Klejman, Laurence and Florence Rochefort, eds. *L'égalité en marche: Le Féminisme sous la Troisième République*. Paris: Presses de la Fondation nationale des sciences politiques, 1989.

Knobel, Marc. "Antisémitisme et ethno-racisme de George Montandon dans les années 1930 et 1940," at http://www.crif.org/uploads/articles/fichiers/Montandon.pdf

– "George Montandon et l'ethno-racisme." In Pierre-André Taguieff, Grégoire Kaufmann, and Michaël Lenoire, eds. *L'Antisémitisme de plume, 1940–1944*, 277–93. Paris: Berg International, 1999.

Kovalewsky, Maxime. Discussion sur Roberty, "Les Préjugés de la sociologie contemporaine." *Annales de l'Institut international de sociologie 1900* 7 (1901): 268–70.

Labarthe, Jean-François-Charles de. "Aperçu général de la science ethnologique." *Revue orientale et américaine* 6 (1861): 342–54.

– "De la Colonisation de Madagascar." *Revue orientale et américaine* 3 (1859–60): 297–303.

– "Esquisse d'un tableau préparatoire génésiaque, pour l'établissement scientifique de l'ethnographie." *Actes de la Société d'ethnographie* 5 (1864–66): 146–57.

– "Essai sur les classements bibliographiques des ouvrages relatifs à l'ethnographie." *Actes de la Société d'ethnographie* 3 (1862–4): 8–9

– "La Menstruation dans les races différentes et des rites de puberté." *Mémoires de la Société d'ethnographie* series 2, 1, or 11 (1869): 67–73, 302–10.

Lacombe, Paul. "Milieu et race." *Revue de synthèse* 2 (1901): 34–55.

Lacombe, Robert. "Essai sur les origines et les premiers développements de la Société d'Ethnographie." *L'Ethnographie* 83 (1980): 329–41.

Lagneau, Gustave. "Influence des milieux sur la race, Modifications mésologiques des caractères ethniques de notre population." *Bulletins de la Société d'anthropologie de Paris* 4th series, 6 (1895): 143–7.

Lamarck, Jean-Baptiste. *Philosophie zoologique*. Paris: Dentu, 1809.

Lambert, Alfred. Review of L[ouis] Brunel, Ch[arles] Mourey, and [Paul] Chemin-Dupontès. "L'Année coloniale." *Revue internationale de sociologie* 14 (1906): 169.

– Review of Finot, "Le Préjugé et le problème des sexes." *Revue internationale de sociologie* 21 (1913): 738.

– Review of Paul Grimanelli, "La femme et le positivisme." *Revue internationale de sociologie* 15 (1907): 391–2.

– Review of Joran, "Autour du féminisme." *Revue internationale de sociologie* 15 (1907): 819.
– Review of Letourneau, "La Condition de la femme dans les diverses races et civilisations." *Revue internationale de sociologie* 11 (1903): 327–30.
– Review of Novicow, "L'affranchissement de la femme." *Revue internationale de sociologie* 11 (1903): 339–41.
– Review of Léopold de Saussure, "Psychologie de la colonisation française." *Revue internationale de sociologie* 7 (1899): 553–4.
Landes, Joan. *Women and the Public Sphere in the Age of the French Revolution.* Ithaca: Cornell University Press, 1988.
Larbiou, Benoît. "Le Corps médical et la race en 1930. Les usages médicaux du racialisme," 111–28. In Carole Reynaud Paligot, ed. *Tous les hommes sont-ils égaux? Histoire comparée des pensées raciales 1860–1930.* Munich: Oldenbourg, 2009.
– "René Martial, 1873–1955. De l'hygiénisme à la raciologie, une trajectoire possible." *Genèses* no. 60 (2005): 98–120.
Larguier de Bancels, Jean. "L'oeuvre d'Alfred Binet." Reprinted in *Centenaire Th. Ribot, Jubilé de la psychologie scientifique française 1839–1889–1939*, 151–66. Agen: Imprimerie moderne, 1939.
Laurent, Goulven, ed. *Jean-Baptiste Lamarck 1744–1829.* Paris: Editions du CTHS, 1997.
Laurière, Christine. *Paul Rivet, Le savant et le politique.* Paris: Publications scientifiques du Muséum national d'histoire naturelle, 2008.
Le Bon, Gustave. "Applications de la psychologie à la classification des races." *Revue philosophique* 22 (1886): 593–619.
– "Correspondance." *Revue d'anthropologie* 2nd series 5 (1882): 747–8.
– "Ethnographie, Congrès international pour l'étude des questions coloniales." *Revue scientifique* 26, 6 (1889): 225–37.
– *Lois psychologiques de l'évolution des peuples* [1894]. Paris: Les amis de Gustave Le Bon, 1978.
– "Notes sur l'étude du caractère." *Revue philosophique* 4 (1877): 496–512.
– "Recherches anatomiques et mathématiques sur les variations du volume du cerveau et sur les relations avec l'intelligence." *Revue d'anthropologie* 8 (1879): 27–104.
– "Recherches expérimentales sur les variations de volume du crâne." *Revue philosophique* 6 (1878): 207–9.
Lebovics, Herman. *True France: The Wars over Cultural Identity, 1900–1945.* Ithaca: Cornell University Press, 1992.
Le Dantec, Félix. "Les Néo-Darwiniens et l'hérédité des caractères acquis." *Revue philosophique* 47 (1899): 1–41.
– Review of Lucien Cuénot, "La Genèse des espèces animales." *Revue philosophique* 72 (1911): 637–42.

– Review of Hugo de Vries, "Espèces et variétés." *Revue philosophique* 67 (1909): 529–32.
– "La Théorie biochimique de l'hérédité." *Revue philosophique* 47 (1899): 457–94.
Lee, Alice, Marie Lewenz, and Karl Pearson. "On the Correlation of the Mental and Physical Characters in Man, Part II." *Proceedings of the Royal Society of London* 71 (1902–3): 106–14.
Lefèvre, André. "Races, peuples, langues de l'Afrique." *Revue mensuelle de l'école d'anthropologie* 2 (1892): 65–79.
Lefèvre, Wolfgang. "Inheritance of Acquired Characters in Lamarck's and Geoffroy Saint-Hilaire's Zoology," 93–107. In Hans-Jörg Rheinberger and Staffan Müller-Wille, eds. *A Cultural History of Heredity* II: *18th and 19th Centuries.* Berlin: Max Planck Institute for the History of Science, 2003.
Lehmann, Jennifer. "The Question of Caste in Modern Society: Durkheim's Contradictory Theories of Race, Class, and Sex." *American Sociological Review* 60 (1995): 566–85.
Lenoir, Raymond. "Lettres de Théodule Ribot à Alfred Espinas (1876–1893)." *Revue philosophique* 165 (1975): 157–72.
Lesonn, Loïg. "Les déboires d'enquête d'Alfred Binet auprès des enfants des écoles primaires de Paris 1897–1905." Communication au colloque de Limoges, 7–8 avril 2005 (Association Française de Sociologie) – RTF 27. Retrieved 31 July 2007 from http://perso.orange.fr/koebel/afs/resumesL.htm#T08
L'Estoile, Benoît de. "'Des races non pas inférieures, mais différentes': de l'exposition coloniale au Musée de l'homme," 391–473. In Claude Blanckaert, *Les Politiques de l'anthropologie.* Paris: L'Harmattan, 2001.
Letourneau, Charles. Review of Le Bon, "L'Homme et la société." *Revue d'anthropologie* 2nd series, 4 (1881): 332–6.
– *La Condition de la femme dans les diverses races et civilisations.* Georges Papillault and Léonce Manouvrier, eds, Paris: V. Giard et E. Brière, 1903.
– "Sur les Phases sociales." *Bulletins de la Société d'anthropologie de Paris* 2nd series, 2 (1867): 378–88.
– *La Psychologie ethnique.* Paris: Schleicher Frères, 1901.
– 'Rapport sur le prix Godard, Review of Le Bon, "Recherches anatomiques." *Bulletins de la Société d'anthropologie de Paris* 3rd series, 2 (1879): 373–86.
– *Science et matérialisme.* 2nd ed. Paris: Reinwald, 1891.
– *La Sociologie d'après l'ethnographie* [1880] 3rd ed. Paris: C. Reinwald, 1892.
Lewontin, Richard A. *Biology as Ideology: The Doctrine of DNA.* Toronto: Anansi, 1991.
– "Race and Intelligence." *Bulletin of the Atomic Scientists* 26 (1970): 2–8.
– *Triple Helix: Gene, Organism, and Environment.* Cambridge, Mass.: Harvard University Press, 2000.

Limousin, Charles. Discussion of La Grasserie, "La race et le milieu comme facteurs d'évolution." *Revue internationale de sociologie* 10 (1902): 392–3.

– Discussion of Roberty, "Les Préjugés de la sociologie contemporaine." *Annales de l'Institut international de sociologie 1900* 7 (1901): 259–67.

– "Le Progrès inégal des races." *Annales de l'Institut international de sociologie 1895* 2 (1896): 261–7.

Logue, William. *From Philosophy to Sociology, The Evolution of French Liberalism, 1870–1914.* De Kalb, Illinois: Northern Illinois University Press, 1983.

– "Sociologie et politique: le libéralisme de Célestin Bouglé." *Revue française de sociologie* 20 (1979): 141–61.

Lopéz-Beltran, Carlos. "In the Cradle of Heredity; French Physicians and *L'Hérédité Naturelle* in the Early 19th Century." *Journal of the History of Biology* 37 (2004): 39–72.

– "Forging Heredity: From Metaphor to Cause, a Reification Story." *Studies in History and Philosophy of Science* 25, 2 (1994): 211–35.

Lorcin, Patricia. *Imperial Identities: Stereotyping, Prejudice and Race in Colonial Algeria.* London: I. B. Tauris, 1995.

Loria, Achille. "Féminisme au point de vue sociologique." *Revue internationale de sociologie* 15 (1907): 5–17.

Lucas, Prosper. *Traité philosophique et physiologique de l'hérédité naturelle dans les états de santé et de maladie du systéme nerveux.* 2 vols. Paris: J.-B. Baillière, 1847–1850.

Lukes, Steven. *Emile Durkheim: His Life and Work, A Historical and Critical Study.* Stanford: Stanford University Press, [1972] 1985.

Macauliffe, Léon. "Variation de la pigmentation des yeux et des cheveux chez les peuples latins." *L'Ethnographie* n.s. 5 (1922): 31–40.

Madier de Montjau, Edouard. *De l'Emigration des Chinois au point de vue des intérêts européens.* Paris: Maisonneuve, 1873.

Maguerie, Jeanne de. Discussion of Ponsinet and La Grasserie, "Le Rôle social de la femme." *Revue internationale de sociologie* 14 (1906): 139–40.

Mahoudeau, Pierre-G. Review of Léonce Manouvrier, "Sur l'Interprétation de la quantité de l'encéphale et du poids de cerveau en particulier." *Revue philosophique* 24 (1887): 317–21.

Malapert, Paulin. Review of Fouillée, "Esquisse psychologique des peuples européens." *Revue philosophique* 55 (1903): 438–41.

Manias, Chris. "The *Race prussienne* Controversy: Scientific Internationalism and the Nation." *Isis* 100 (2009): 733–57.

Manouvrier, Léonce. "L'Anthropologie et le droit." *Revue internationale de sociologie* 2 (1894): 241–73, 351–70.

– "Anthropométrie et aptitudes." *Revue anthropologique* 21 (1911): 409–29, 475–87.

– "Les Aptitudes et les actes dans leurs rapports avec la constitution anatomique et avec le milieu extérieur." *Bulletins de la Société d'anthropologie de Paris* 4th series, 1 (1890): 918–51.
– "L'Atavisme et le crime." *Revue mensuelle de l'Ecole d'anthropologie* 1 (1891): 225–40.
– "Cerveau." In Charles Richet, ed. *Dictionnaire de physiologie* 12: 670–745. Paris: Félix Alcan, 1897.
– "Le Classement universitaire de l'anthropologie." *Revue de l'Ecole d'anthropologie* 17 (1907): 75–96, 109–19.
– "Conclusions générales sur l'anthropologie des sexes." *Revue de l'Ecole d'anthropologie* 19 (1909): 41–61.
– "Cours d'ethnologie, L'étude comparative des sexes dans les races humaines." *Le Progrès français.* 21 novembre 1883: 1–4.
– Discussion of Mécislas Golberg, "L'Origine des races et la division du travail." *Annales de l'Institut international de sociologie 1895* 2 (1896): 369–85.
– "Essai sur les qualités intellectuelles considérées en fonction de la superiorité cérébrale quantitative." *Revue mensuelle de l'Ecole d'anthropologie* 4 (1894): 65–84.
– "L'Ethnologie et l'Ethnographie dans l'Anthropologie." *L'Homme* 1 (1884): 161–71.
– "La Genèse normale du crime." *Bulletins de la Société d'anthropologie de Paris* 4th series, 4 (1893): 405–58.
– "L'Indice céphalique et la pseudo-sociologie." *Revue mensuelle de l'Ecole d'anthropologie* 9 (1899): 233–59, 280–96.
– "Le Progrès anthropologique de l'intelligence." *Annales de l'Institut international de sociologie 1912* 4 (1913): 77–100.
– "Quelques cas de criminalité juvénile." *Revue anthropologique* 22 (1912): 425–38.
– "Rapports de la sociologie et de l'anthropologie." *Revue internationale de sociologie* 12 (1904): 877–8.
– Review of Anténor Firmin, "L'Egalité des races humaines." *Revue philosophique* 21 (1886): 180–2.
– Review of Topinard, "Science et foi." *Revue de l'Ecole d'anthropologie* 10 (1900): 210–12.
Marchal, Anne and Serge Nicolas, "L'emancipation de la psychologie à travers trois textes majeurs de Ribot." *Psychologie et Histoire* 1 (2000): 43–81. http://psychologieethistoire.googlepages.com/MARCHAL1.HTM
Marie, Auguste. "Notes sur la morphologie de l'encéphale des Asiatiques." *L'Ethnographie* new series, 1, no. 2 (1914): 81–8
– "Variation de la pigmentation des yeux et des cheveux chez les peuples latins." *L'Ethnographie* n.s. 5 (1922): 41–8
Marie, Auguste and Léon MacAuliffe. "L'Anthropométrie des Malgaches." *L'Ethnographie* no. 3 (1914): 71–2.

Marin, Louis. Compte-rendu des séances, *L'Ethnographie* n.s. 5 (1922): 10–11.

Marion, Henri. Review of Armand de Quatrefages, "L'Espèce humaine." *Revue philosophique* 3 (1877): 627–43.

Martial, Lydie, Jacques Novicow, Raoul de La Grasserie, Paul Vibert, and René Worms. Discussion of Moll-Weiss, "La Ménagère et le féminisme." *Revue internationale de sociologie* 18 (1910):507–15.

Martial, René. "Anthropo-biologie des races, L'Esprit sous la matière." *Mercure de France* 288 (1 décembre 1938): 276–303.

– "Etrangers et métis." *Mercure de France* 294 (15 septembre 1939): 513–27.

– "L'immigration et l'avenir de la France." *Mercure de France* 245, no. 842 (15 juillet 1933): 257–92.

– *Les Métis.* Paris: Flammarion, 1942.

– "Politique de l'immigration." *Mercure de France* 259, no. 884 (15 avril 1935): 267–94.

– "Un Précurseur des groupements sanguins: Vacher de Lapouge." *Mercure de France* 270 (15 septembre 1936): 620–5.

– *La Race française.* Paris: Mercure de France, 1934.

– *Vie et constance des races.* Paris: Mercure de France, 1939.

Martinet, Ludovic. Review of Arthur Bordier, "La Colonisation scientifique." *Revue d'anthropologie* 7 (1884): 330–1.

Massin, Benoît. "L'anthropologie raciale comme fondement de la science politique: Vacher de Lapouge et l'échec de 'l'anthroposociologie' en France 1866–1936," 269–334. In Blanckaert, ed. *Les politiques de l'anthropologie.* Paris: L'Harmattan, 2001.

Mauger, Gérard. "Les Fossoyeurs d'anthropologie." *L'Ethnie française* no. 1 (mars 1941): 14–15.

Maunier, René. Review of Ghénia Avril de Sainte-Croix, "Le Féminisme." *Revue internationale de sociologie* 15 (1907): 667.

Mauss, Marcel. *Oeuvres.* 3 vols. Paris: Editions du minuit, 1969.

Mauxion, Marcel. "Les Eléments et l'évolution de la moralité." *Revue philosophique* 56 (1903): 1–29, 150–80.

McMillan, James. *France and Women, 1789–1914: Gender, Society, and Politics.* London: Routledge, 2000.

Meaney, Michael J. "Nature, Nurture, and the Disunity of Knowledge." *Annals of the New York Academy of Sciences* 935 (2001): 50–61.

Mémoires de la Société d'anthropologie de Paris.

Mémoires de la Société d'ethnographie.

Mémoires de la Société ethnologique de Paris, t. 1. Paris: Librairie orientale de Mme Ve Dondey-Dupré 1841.

Mémoires et travaux de la Société d'ethnographie.

Meyran, Régis. "Vichy ou la face cachée de la République." *L'Homme* no. 160 (2001): 177–84.

Misme, Jane. "La Femme de lettres." *Revue internationale de sociologie* 17 (1909): 14–30.
Moll-Weiss, Augusta. "La Ménagère et le féminisme." *Revue internationale de sociologie* 18 (1910): 499–507 (including discussion).
Monin, Hippolyte. Discussion of Golberg, "L'Origine des races et la division du travail." *Annales de l'Institut international de sociologie 1895* 2 (1896): 372–3.
Montandon, George. "L'Aryanisme français." *L'Ethnie francaise* no. 2 (avril 1941): 2–6.
– "Les Caractères du type judaïque." *L'Ethnie française* no. 3 (mai-juin 1941): 15–20.
– *L'Ethnie française.* Paris: Payot, 1935.
– "Ethno-raciologie judaïque." *L'Ethnie française* no. 8 (mai 1943): 5–9.
– "L'Etudiant français et la science ethnique." *L'Ethnie française* no. 6 (mars 1942): 2–6.
– " Georges Vacher de Lapouge." *L'Ethnie française* 3 (mai-juin 1941): 3–7.
– *La Race, les races.* Paris: Payot, 1933.
– "La Religion juive, signe de ralliement ethnique." *L'Ethnie française* no. 5 (août-september 1941): 25–6.
Moreau de Tours, Jacques. *La Psychologie morbide dans ses rapports avec la philosophie de l'histoire.* Paris: Victor Masson, 1859.
Morel, Bénédict-Augustin. *Traité des dégénérescences physiques, intellectuelles et morales de l'espèce humaine.* Paris: J.-B. Baillière, 1857.
Mortillet, Gabriel. "Anti-Sémitisme." *L'Homme* 1 (1884): 522–8; 4 (1887): 606–7.
Mosbah-Natanson, Sébastien. "Internationalisme et tradition nationale: le cas de la constitution de la sociologie française autour de 1900." *Revue d'histoire des sciences humaines* 18 (2008): 35–62.
Mougniotte, Alain. "Alfred Binet et 'la chose politique'." *Cahiers Binet-Simon* 634 (1993): 41–9.
Mucchielli, Laurent. "Autour des 'Instructions aux voyageurs sur les Boschimans' d'Henri Thulié; méthodes, enjeux, et conflits de l'anthropologie française à la fin du XIXe siècle," 201–42. In Claude Blanckaert, ed. *Le Terrain des sciences humaines.* Paris: L'Harmattan, 1996.
– "La Dénaturalisation de l'homme: Le tournant Durkheimien de l'ethnologie française (1890–1914)," 41–53. In Albert Ducros, Jacqueline Ducros, and Frédéric Joulian, eds. *La culture est-elle naturelle?* Paris: Errance, 1998.
– *La Découverte du social, naissance de la sociologie en France.* Paris: La Découverte, 1998.
– "Hérédité et 'Milieu social,' le faux antagonisme franco-italien, la place de l'Ecole de Lacassagne dans l'histoire de la criminologie," 189–214. In Laurent Mucchielli, ed. *Histoire de la Criminologie française.* Paris: L'Harmattan, 1994.
– "Aux Origines de la psychologie universitaire en France 1870–1900: enjeux intellectuels, contexte politique, réseaux et stratégies d'alliance autour de la

'Revue Philosophique' de Théodule Ribot." *Annals of Science* 55 (1998): 263–89.
– "Psychologie des peuples, races, regions et milieu social. Problèmes scientifiques et enjeux disciplinaires d'une théorie de l'histoire autour d'Henri Berr et de la *Revue de synthèse historique* (1890–1925)," 81–100. In Agnès Biard, Domique Bourel, et Eric Brian, *Henri Berr et la Culture du XXe siècle. Histoire, science et philosophie.* Paris: Albin Michel/Centre international de synthèse, 1997.
– "Psychologie et sociologie en France, l'appel à un territoire commun: vers une psychologie collective (1890–1940)." *Revue de synthèse* 3–4 (1994): 445–83.
– "Sociologie versus anthropologie raciale. L'engagement des sociologues durkheimiens dans le contexte 'fin de siècle' (1885–1914)." *Gradhiva* 21 (1997): 77–95.
Mucchielli, Laurent and Daniel Becquemont, *Le Cas Spencer.* Paris: PUF, 1998.
Muel-Dreyfus, Francine. *Vichy and the Eternal Feminine: A Contribution to a Political Sociology of Gender* [1996]. Durham: Duke University Press, 2001.
Muffang, Henri. "Etudes d'anthropo-sociologie: écoliers et paysans de Saint-Brieuc." *Revue internationale de sociologie* 5 (1897): 789–803.
Nadaillac, Jean-François-Albert du Pouget, marquis de. "Les Esquimaux." *L'Anthropologie* 13 (1902): 94–104.
Nicolas, Serge. "Alfred Binet 1857–1911: Aperçus sur sa vie et son oeuvre," 15–28. In Bernard Andrieu, ed. *Oeuvres complètes d'Alfred Binet.* St Pierre du Mont: Euredit, 2001.
– "Alfred Binet et 'l'Année psychologique' d'après une correspondance inédite." *L'Année psychologique* 97 (1997): 665–99.
– "Le Binet-Simon et les origines de la mesure de l'intelligence," 19–39. In Serge Nicolas and Bernard Andrieu, eds. *La Mesure de l'intelligence.* Paris: L'Harmattan, 2005.
– "L'Hérédité psychologique d'après Théodule Ribot (1873): La Première these de psychologie 'scientifique'." *L'Année psychologique* 99 (1999): 295–348.
– *Histoire de la psychologie.* Paris: Dunod, 2001.
– "L'Introduction de l'enseignement de la psychologie scientifique en France: Théodule Ribot (1839–1916) à la Sorbonne (1885)." *L'Année psychologique* 100 (2000): 285–331.
– *Théodule Ribot, Philosophe breton, fondateur de la psychologie française.* Paris: L'Harmattan, 2005.
Nicolas, Serge and Agnes Charvillat, "Introducing Psychology as an Academic Discipline in France: Théodule Ribot and the Collège de France (1888–1901)." *Journal of the History of the Behavioral Sciences* 37, 2 (2001): 143–64.
Nicolas, Serge and Ludovic Ferrand. "Alfred Binet and Higher Education." *History of Psychology* 5 (2002): 264–83.

Nicolas, Serge, Anne Marchal, and F. Isel. "La psychologie au XIXe siècle." *Revue d'histoire des sciences humaines* 2 (2000): 57–103.

Nicolas, Serge and David Murray, "Le Fondateur de la psychologie 'scientifique' française: Théodule Ribot (1839–1916)." *Psychologie et Histoire* 1 (2000): 1–42. http://psychologieethistoire.googlepages.com/NICOLAS1.HTM

Nicolas, Serge, Juan Ségui, and Ludovic Ferrand. "Les premières revues de psychologie: La place de *l'Année Psychologique.*" *L'Année Psychologique* 100 (2000): 71–110.

Noiriel, Gérard. *Etat, nation et immigration, Vers une histoire du pouvoir.* Paris: Belin 2001

– *The French Melting Pot: Immigration, Citizenship, and National Identity* [1988]. Minneapolis: University of Minnesota Press, 1996.

– *Immigration, antisémitisme, et racisme en France (XIXe-XXe siècle), Discours publics, humiliations privées* [2007]. Paris: Hachette, 2009.

– *Les Origines républicaines de Vichy.* Paris: Hachette, 1999.

Nordmann, Jean-Thomas. *Taine et la critique scientifique.* Paris: PUF, 1992.

Novicow, Jakob Alexandrovich (Jacques). *L'Affranchissement de la femme.* Paris: F. Alcan, 1903.

– "Les Bases biologiques, psychologiques et sociologiques du droit." *Revue philosophique* 70 (1910): 1–20.

– "Les Castes et la sociologie biologique." *Revue philosophique* 50 (1900): 361–73.

– Discussion of Golberg, "L'Origine des races et la division du travail." *Annales de l'Institut international de sociologie 1895* 2 (1896): 368.

– Discussion of Roberty, "Les Préjugés de la sociologie contemporaine." *Annales de l'Institut international de sociologie 1900* 7 (1901): 2712.

– "Le Péril jaune." *Revue internationale de sociologie* 5 (1897): 351–5.

Nye, Robert A. *Crime, Madness, and Politics in Modern France: The Medical Concept of National Decline.* Princeton: Princeton University Press, 1984.

– *The Origins of Crowd Psychology: Gustave Le Bon and the Crisis of Mass Democracy in the Third Republic.* London: Sage, 1975.

Offen, Karen. "Defining Feminism: A Comparative Historical Approach." *Signs* 14, 1 (1988): 119–57; and discussion with Nancy Cott, *Signs* 15, 1 (1989): 203–9.

– "Depopulation, Nationalism, and Feminism in Fin-de-Siècle France." *American Historical Review* 89 (1984): 648–76.

– *European Feminisms, 1700–1950: A Political History.* Stanford: Stanford University Press, 2000.

– "'La plus grande féministe de France': Mais qui est donc Madame Avril de Sainte-Croix?" *Archives du Féminisme,* bulletin no. 9 (Dec. 2005) http://www.archivesdufeminisme.fr/article.php3?id_article=127.

– "Intrepid Crusader: Ghénia Avril de Sainte-Croix Takes on the Prostitution Issue." *Proceedings of the Western Society of French History* 33 (2005): 352–74.

– "Madame Ghénia Avril de Sainte-Croix, the Josephine Butler of France." *Women's History Review* 17, 2 (2008): 239–55.

Orr, David. "Darwinian Storytelling." *New York Review of Books*, 27 February 2003.

Osborne, Michael. *Nature, the Exotic, and the Science of French Colonialism.* Bloomington, Ind.: Indiana University Press, 1994.

Otis, Laura. *Organic Memory: History and the Body in the late 19th and early 20th centuries.* Lincoln, Nebraska: University of Nebraska Press, 1994.

Paligot, Carole Reynaud. "La Psychologie des peuples et ses applications durant l'entre-deux-guerres." *Revue de synthèse* 129 6th series, no. 1 (2008): 125–46.

– *Races, racisme et antiracisme dans les années 1930.* Paris: PUF, 2007.

– *La République raciale, Paradigme racial et idéologie républicaine, 1860–1930.* Paris: PUF, 2006.

Papillault, Georges. "L'Anthropologie est-elle une science unique?" *Revue de l'Ecole d'anthropologie* 18 (1908): 117–32.

– "Bio-sociologie." *Revue anthropologique* 22 (1912): 1–19.

– "Condition d'une enquête scientifique sur les criminels." *Revue anthropologique* 23 (1913): 41–56.

– Discussion of La Grasserie, "La Race et le milieu comme facteurs de l'évolution." *Revue internationale de sociologie* 10 (1902): 387–90.

– "Galton et la biosociologie." *Revue de l'Ecole d'anthropologie* 21 (1911): 56–65.

– "Sur Quelques erreurs de méthode en criminologie." *Revue de l'Ecole d'anthropologie* 20 (1910): 321–34.

– Review of Lucien Cuénot, "La Genèse d'une espèce animale." *Revue de l'Ecole d'anthropologie* 21 (1911): 154.

Papillon, Fernand. Review of Ribot, "Hérédité psychologique." *Revue politique et littéraire* 11 (1873): 1214–18.

Parodi, Dominique. Review of Célestin Bouglé, "Les Idées égalitaires, Etude sociologique." *Revue philosophique* 49 (1900): 545–52.

Paul, Diane. *The Politics of Heredity: Essays on Eugenics, Biomedicine, and the Nature-Nurture Debate.* Albany: SUNY Press, 1998.

Paulhan, Frédéric. Review of Paul Lacombe, "La Psychologie des individus et des sociétés chez Taine historien des litteratures." *Revue philosophique* 62 (1906): 419–25

– Review of Le Bon, "Les Lois psychologiques de l'évolution des peuples." *Revue philosophique* 38 (1894): 513–22.

– Review of Le Bon, "La Psychologie des foules." *Revue philosophique* 41 (1896): 79–85.

– Review of Letourneau, "La Sociologie d'après l'ethnographie." *Revue philosophique* 11 (1881): 546–51.

– Review of Novicow, "La Morale de l'intérêt dans les rapports individuels et internationaux" and "Mécanisme et limites de l'association humaine." *Revue philosophique* 73 (1912): 290–301.

Pedersen, Jean Elizabeth. *Legislating the French Family: Feminism, Theater, and Republican Politics, 1870–1920*. New Brunswick: Rutgers University Press, 2003.
– "Sexual Politics in Comte and Durkheim: Feminism, History, and the French Sociological Tradition." *Signs* 27 (2001): 229–63.
Pick, Daniel. *Faces of Degeneration: A European Disorder, c. 1848–c. 1918*. Cambridge: Cambridge University Press, 1989.
Pinker, Steven. *The Blank Slate: The Modern Denial of Human Nature*. New York: Viking, 2002.
– "A History of Violence," *The New Republic* (19 March 2007). http://www.tnr.com/article/history-violence-were-getting-nicer-every-day
– "The Science of Gender and Science," Harvard University Mind/Brain/Behavior debate with Elizabeth Spelke, (16 May 2005) http://www.edge.org/3rd_culture/debate05/debate05_index.html
Plas, Régine. "La Psychologie pathologique d'Alfred Binet," 229–45. In Paul Fraisse and Juan Ségui, eds. *Les Origines de la psychologie scientifique*. Paris: PUF, 1994.
Policar, Alain. "Science et démocratie: Célestin Bouglé et la métaphysique de l'hérédité." *Vingtième siècle. Revue d'histoire* no. 61 (1999): 86–101.
Ponsinet, L. "Le Rôle de la femme dans le progrès de la civilisation." *Revue internationale de sociologie* 14 (1906): 112–26.
Pournin, Marcel. Review of Paul Leroy-Beaulieu, "De la Colonisation chez les peuples modernes." *Revue internationale de sociologie* 12 (1904): 373–4.
Psychology, Public Policy, and Law 11, 2 (2005).
Quatrefages, Armand de. *Discours d'ouverture au cours d'anthropologie professé au Muséum*, extrait de la *Gazette médicale de Paris*. Paris np, 1861.
– *L'Espèce humaine*. Paris: Félix Alcan, 1896.
– *The Human Species*. London: C. Kegan Paul, 1879.
– *The Human Species*. New York: Appleton, 1879.
– *Rapport sur les progrès de l'anthropologie*. Paris: Imprimerie impériale, 1867.
– *Unité de l'espèce humaine*. Paris: J. Claye, 1861.
Quinlan, Sean. "The Racial Imagery of Degeneration and Depopulation: Georges Vacher de Lapouge and 'Anthroposociology' in Fin-de-siècle France." *History of European Ideas* 24, 6 (1998): 393–413.
Reinach, Salomon. Discours, *Institut français d'anthropologie. Compte-rendu des séances*, supplément to *L'Anthropologie* 22 (1911): 8–43.
– "La prohibition de l'inceste et le sentiment de la pudeur." *L'Anthropologie* 10 (1899): 59–70; 13 (1902): 655–69.
Renan, Ernest. *Qu'est-ce qu'une nation?* Paris: Calmann Lévy, 1882.
Renouvier, Charles. "Des rapports du physique et du moral." *La Critique philosophique* 2 (1873): 202–8, 313–20, 357–63.
Réthoré, François. "Plaidoyer pour le dogmatisme." *Revue orientale et américaine* 2nd series, 7 (1893): 159–61.
Revue d'anthropologie

Revue d'ethnographie 1882–89
Revue géographique
Revue géographique internationale
Revue internationale de sociologie
Revue orientale et américaine
Revue philosophique de la France et de l'étranger
Ribot, Théodule. *English Psychology*. London: Henry A. King, 1873, reprint Thoemmes, 2002.
– *Essai sur l'imagination créatrice*. Paris: Félix Alcan, 1900.
– *Evolution des idées générales*. Paris: Félix Alcan, 1897.
– "L'Hérédité." In A. Berthelot et al. *La Grande Encyclopédie* 19:1165–9. Paris: H. Lamirault, 1894.
– *L'Hérédité, étude psychologique sur ses phénomènes, ses lois, ses causes, ses conséquences*. Paris: Ladrange, 1873.
– *L'Hérédité psychologique*. Paris: Félix Alcan, 1894.
– *La Logique des sentiments* [1905]. Paris: L'Harmattan, 1998.
– "Les Méthodes de la psychologie." In F. Thomas, ed. *De la Méthode dans les sciences*. Paris: Alcan, 1909, republished (2000) in A. Marchal and S. Nicolas, "L'Emancipation de la psychologie vue à travers trios textes majeurs de Ribot," *Psychologie et histoire* 1 (2000): 43–81. http://psychologieethistoire.googlepages.com/MARCHAL1.HTM
– *La Psychologie allemande contemporaine*. Paris: Germer Baillière, 1879.
– *La Psychologie anglaise contemporaine* [1870]. Paris: Germer Baillière, 1881.
– "La Psychologie ethnographique en Allemagne." *Revue philosophique* 2 (1876): 596–607.
– *La Psychologie des sentiments*. Paris: Félix Alcan, 1896.
– Review of Candolle, "Histoire des sciences et des savants." *Revue philosophique* 19 (1885): 328–31.
– "M. Taine et sa psychologie." *Revue philosophique* 4 (1877): 17–46.
– "De la Valeur des questionnaires en psychologie." *Journal de psychologie normale et pathologique* 1 (1904): 1–10.
Richard, Gaston. "Les Lois de la solidarité morale." *Revue philosophique* 60 (1905): 441–71.
– Review of Letourneau, "L'Evolution de l'esclavage." *Revue philosophique* 43 (1897): 643–51.
– Review of Letourneau, "La Condition de la femme dans les diverses races et civilisations." *Revue philosophique* 56 (1903): 429–33.
– Review of Ernest Seillière, "Le Comte de Gobineau et l'Aryanisme historique." *Revue philosophique* 56 (1903): 433–6.
– Review of Jacques Lourbet, "La Femme devant la science contemporaine." *Revue philosophique* 43 (1897): 434–5.

Richard, Nathalie. "La Revue *L'Homme* de Gabriel de Mortillet: anthropologie et politique au début de la Troisième République." *Bulletins et mémoires, Société d'anthropologie de Paris* new series, 1, no. 3–4 (1989): 221–56.

Richet, Charles. "Rapport sur un projet de questionnaire d'hérédité psychologique." *Bulletins de la Société d'anthropologie de Paris* 3rd series, 7 (1884): 734–40.

Ridley, Matt. *Nature via Nurture: Genes, Experience, and What Makes Us Human.* Toronto: Harper Collins, 2003.

Rignol, Loïc. "Anthropologie et progrès dans la philosophie de l'histoire d'Alphonse Esquiros, Le système des *Fastes populaires.*" *Revue d'histoire du XIXe siècle* n° 20–21 (2000): 157–79.

Rignol, Loïc and Philippe Régnier, "Races et politiques dans l'Histoire de France chez Victor Courtet de l'Isle (1813–1867): Enjeux de savoir et luttes de pouvoir au XIXe siècle," 127–52. In Philippe Régnier, ed. *Etudes Saint-Simoniennes.* Lyon: Presses universitaires de Lyon, 2002.

Rivet, Paul. "Le Christianisme et les Indiens de la République de l'Equateur." *L'Anthropologie* 17 (1906): 81–101.

– "Etude sur les Indiens de la région de Riobamba." *Journal de la Société des américanistes* nouvelle série, 1, 1 (1903): 58–80.

– "Les Indiens Colorados, récit de voyage et étude ethnologique." *Journal de la Société des américanistes* 2, 1 (1905): 177–208.

– "Les Indiens Jibaros, Etude géographique, historique, ethnographique." *L'Anthropologie* 18 (1907): 333–68, 538–618; 19 (1908): 69–87, 235–59.

– "Les Indiens de Mallasquer, étude ethnologique." *Bulletins et mémoires de la Société d'anthropologie de Paris* 5th series, 5 (1904): 144–52.

– Review of Hervé, "Noir et blanc, Le Croisement des races aux Etats-Unis et la théorie de la 'miscégénation.' *L'Anthropologie* 18 (1907): 444–5.

Roberts, Mary Louise. *Disruptive Acts: The New Woman in Fin-de-Siècle France.* Chicago: University of Chicago Press, 2002.

Roberty, Eugène. "Les Préjugés de la sociologie contemporaine."*Annales de l'Institut international de sociologie 1900* 7 (1901): 239–59.

Rochefort, Florence. "The French Feminist Movement and Republicanism, 1868–1914," 77–101. In Sylvia Paletschek and Bianka Pietrow-Ennker, eds. *Women's Emancipation Movements in the Nineteenth Century: A European Perspective.* Stanford: Stanford University Press, 2004.

Rosny, Léon. "L'Amérique." *Comptes rendus des séances de la Société d'ethnographie* 1 (1860): 99–117.

– "Eléments d'une ethnographie descriptive." *Actes de la Société d'ethnographie* 6 (1867–70): 204–21.

– "Essai d'ethnographie théorique et appliquée," *Mémoires de la Société d'ethnographie* 2nd series, 1 (1869): 5–24.

– "L'Evolution politique et sociale du Japon contemporain." *Bulletin de la Société d'ethnographie* 2nd series, 7 (1893–4): 208–16.
– "L'Idée de l'égalité chez les sociétés et chez les individus." *Bulletin de la Société d'ethnographie* 2nd series, 3 or 13 (1889): 16–29.
– "Instructions ethnographiques." *Actes de la Société d'ethnographie* 5 (1864–6): 26–9
– *La Méthode conscientielle: Essai de philosophie exactiviste.* Paris: Félix Alcan, 1887.
– "De la Méthode ethnographique." *Actes de la Société d'ethnographie* 7 (1871–73): 41–60.
– "L'Orient." *Comptes rendus des séances de la Société d'ethnographie* 1 (1860): 48–59.
– *Rapport annuel fait à la S. d'Ethnog. Sur ses travaux et sur les progrès des sciences ethnographiques pendant l'année 1864.* Paris: Challamel aîné, 1865.
– *Traité d'ethnographie théorique et descriptive.* 2 vols. Paris: Ernest Leroux, 1900–1902.
Royer, Clémence. "Deux hypothèses sur l'hérédité." *Revue d'anthropologie* 6 (1877): 443–84.
Saada, Emmanuelle. "Race and Sociological Reason in the Republic: Inquiries on the Métis in the French Empire (1908–37)." *International Sociology* 17, 3 (2002): 361–91.
Sabatier, Camille. "De l'Art du peuple éducateur." *L'Homme* 4 (1887): 33–46.
– "Les Berbères sédentaires." *Revue d'anthropologie* 2nd series, 5 (1882): 412–42.
– "Essai sur l'ethnologie de l'Afrique du nord." *Revue d'anthropologie* 2nd series, 7 (1884): 404–59.
– "Etude sur la femme Kabyle." *Revue d'anthropologie* 2nd series, 6 (1883): 56–69.
– "De l'Evolution du progrès dans les groupes sociaux." *L'Homme* 3 (1886): 417–29.
– "Du Progrès indéfini de l'humanité." *L'Homme* 3 (1886): 321–34.
Schiller, Francis. *Paul Broca, Founder of Anthropology, Explorer of the Brain.* [1979] Oxford: Oxford University Press 1992
Schneider, William H. *An Empire for the Masses: The French Popular Image of Africa, 1870–1900.* Westport, Conn.: Greenwood Press, 1982.
– "Hérédité, sang et opposition à l'immigration dans la France des années trente." *Ethnologie française* 24, 1 (1994): 104–17.
– *Quality and Quantity: The Quest for Biological Regeneration in Twentieth-Century France.* Cambridge: Cambridge University Press, 1990.
Schrader, Franz. "L'Afrique du nord maghrébin." *Revue anthropologique* 23 (1913): 76–88.
– "Géographie anthropologique." *Revue mensuelle de l'école d'anthropologie* 3 (1893): 219–53; 7 (1897): 129–41; 10 (1900): 122–5.

– "Les Lois de préservation aux Etats-Unis." *Revue anthropologique* 24 (1914): 1–19

Schuerkens, Ulrike. "Les Congrès de l'Institut international de sociologie de 1894 à 1930 et l'internationalisation de la sociologie." *International Review of Sociology* new series 6, 1 (1996): 7–24.

Scott, Joan Wallach. *Only Paradoxes to Offer.* Cambridge, Mass.: Harvard University Press, 1996.

– *Parité: Sexual Equality and the Crisis of French Universalism.* Chicago: University of Chicago Press, 2005

Seignobos, Charles. Review of Le Bon, "Les Premières civilisations." *Revue philosophique* 28 (1889): 197–202.

Semlali, Yusef. "Eugène de Roberty (1843–1915). Une page peu connue de l'histoire de la sociologie." 2005. http://halshs.archives-ouvertes.fr/halshs-00003964/en/

Shenk, David. *The Genius in All of Us: Why Everything You've Been Told About Genetics, Talent, and IQ is Wrong.* New York: Doubleday, 2010.

Sherman, Daniel J. "'Peoples Ethnographic': Objects, Museums, and the Colonial Inheritance of French Ethnology." *French Historical Studies* 27, 3 (2004): 669–703.

Shields, James. *The Extreme Right in France: From Pétain to Le Pen.* London: Routledge, 2007.

Sibeud, Emmanuelle. "Un Ethnographe face à la colonisation: Arnold van Gennep en Algérie (1911–1912)." *Revue d'histoire des sciences humaines* 10 (2004): 79–104.

– "The Metamorphosis of Ethnology in France, 1839–1930." 96–110. In Henrika Kuklick, ed. *A New History of Anthropology.* Oxford: Blackwell, 2008.

– "'Negrophilia,' 'Negrology' or 'Africanism'? Colonial Ethnography and Racism in France around 1900," 156–67. In Tony Chafer and Amanda Sackur, eds. *Promoting the Colonial Idea: Propaganda and Visions of Empire in France.* Basingstoke: Palgrave, 2002.

– "Projet de présentation d'un bureau d'ethnologie (1913)." *Revue d'histoire des sciences humaines* 10 (2004): 105–15.

– *Une Science impériale pour l'Afrique? La construction des savoirs africanistes en France, 1878–1930.* Paris: Editions de l'EHESS, 2002.

Silva Telles, chevalier de. "La Dégenerescence des races humaines," Congrès international d'anthropologie et d'archéologie préhistoriques. *L'Anthropologie* 13 (1902): 241–3.

Simon, Théodore. *Documents relatifs à la corrélation entre le développement physique et la capacité intellectuelle.* Paris: G. Carré et C. Naud, 1900.

"Société d'anthropologie de Paris, Enquête sur les croisements ethniques." *Revue anthropologique* 22 (1912): 337–407.

Société de sociologie. "Le Colonial, type sociologique." *Revue internationale de sociologie* 18 (1910): 114–39.

– "L'Hérédité des professions." *Revue internationale de sociologie* 7 (1899): 50–9, 117–24, 206–7.

Souriau, Paul. Review of Fouillée, "Tempérament et caractère selon les individus, les sexes, et les races." *Revue philosophique* 40 (1895): 624–33.

Sowerwine, Charles. "The Sexual Contract(s) of the Third Republic," 245–53. In Ian Coller, Helen Davies, and Julie Kalman, eds. *French History and Civilization. Papers from the George Rudé Seminar.* Melbourne: The George Rudé Society, 2005.

– *Sisters or Citizens? Women and Socialism in France since 1876.* Cambridge: Cambridge University Press, 1982.

– "Woman's Brain, Man's Brain: Feminism and Anthropology in Late Nineteenth-Century France." *Women's History Review* 12, 3 (2003): 289–307.

Sparks, Corey S. and Richard L. Jantz, "A reassessment of human cranial plasticity: Boas revisited." *Proceedings of the National Academy of Sciences* 99 (2002): 14636–9.

Spencer, Herbert. "Esquisse d'une psychologie comparée de l'homme." *Revue philosophique* 1 (1876): 45–61.

– *Principes de psychologie*, Théodule Ribot and Alfred Espinas, trans. 2 vols. Paris: Félix Alcan, 1892.

Staum, Martin S. *Cabanis: Enlightenment and Medical Philosophy in the French Revolution.* Princeton: Princeton University Press, 1980.

– *Labeling People: French Scholars on Society, Race, and Empire, 1815–1848.* Montreal: McGill-Queen's, University Press, 2003.

– "Paris Ethnologists and the Perfectibility of 'Races.'" *Canadian Journal of History* 35, 3 (2000): 453–72.

Stepan, Nancy. *The Idea of Race in Science: Great Britain 1800–1960.* London: Macmillan, 1982.

– "Race and Gender: The Role of Analogy in Science." *Isis* 77 (1986): 261–77.

– "Race, Gender, Science, and Citizenship." *Gender and History* 10 (1998): 26–52.

Stiles, Joan. *The Fundamentals of Brain Development: Integrating Nature and Nurture.* Cambridge, Mass.: Harvard University Press, 2008.

Stocking, Jr, George W. "Qu'est-ce qui est en jeu dans un nom?: La Société Ethnographique et l'historiographie de l''anthropologie' en France," 421–32. In Britta Rupp-Eisenreich, ed. *Histoires de l'anthropologie: XVIe - XIXe siècles.* Paris: Klincksieck, 1984.

Stoler, Ann."Sexual Affronts and Racial Frontiers: European Identities and the Cultural Politics of Exclusion in Southeast Asia," 198–237. In Frederick Cooper and Ann Stoler, eds. *Tensions of Empire: Colonial Cultures in a Bourgeois World.* Berkeley: University of California Press, 1997.

Surkis, Judith. *Sexing the Citizen: Morality and Masculinity in France, 1870–1920.* Ithaca: Cornell University Press, 2006.

Sweeney, Gerald. *Fighting for the Good Cause: Reflections on Francis Galton's Legacy to American Hereditarian Psychology.* Philadelphia: American Philosophical Society, 2001.

Sylvain, Benito. *Du Sort des indigènes dans les colonies d'exploitation.* Paris: L Boyer, 1901.

Sylvain, Georges. Discussion of Louis Skarzynski, "Les Peaux-Rouges et les Nègres aux Etats-Unis." *Revue internationale de sociologie* 18 (1910): 362.

Taguieff, Pierre-André. "L'Antisémitisme à l'époque de Vichy, La haine, la lettre et la loi," 45–143. In Pierre-André Taguieff, Grégoire Kauffmann, and Michael Lenoire, eds. *L'antisémitisme de plume 1940–1944, Etudes et documents.* Paris: Berg International,1999.

– *La Couleur et le sang: doctrines racistes à la française.* Paris: Editions mille et une nuits, 2002.

– "Face à l'immigration: Mixophobie, xénophobie ou sélection. Un débat français dans l'entre-deux-guerres." *Vingtième Siècle. Revue d'histoire* no. 47 (Jul. – Sep. 1995): 103–31.

– *La Force du préjugé, Essai sur le racisme et ses doubles.* Paris: Gallimard, 1987.

– "Immigrés, métis, Juifs. les raisons de l'inassimilabilité, Opinions et doctrines du Dr. Martial," 177–220. In Liliane Crips, ed. *Nationalismes, féminismes, exclusions: mélanges en l'honneur de Rita Thalmann.* Berne: Peter Lang, 1994.

– "La 'Science' du Docteur Martial," 295–332. In Pierre-André Taguieff, Grégoire Kauffmann, and Michael Lenoire, eds. *L'antisémitisme de plume 1940–1944, Etudes et documents.* Paris: Berg International, 1999.

Taine, Hippolyte. *Histoire de la littérature anglaise.* 4 vols. Paris: L. Hachette, 1863–4.

– "Notes sur l'acquisition du langage chez les enfants et dans l'espèce humaine," *Revue philosophique* 1 (1876): 5–23.

Tarde, Gabriel. "Criminalité et santé sociale." *Revue philosophique* 39 (1895): 148–62.

– "La Logique sociale des sentiments." *Revue philosophique* 36 (1893): 561–94.

– "La Misère et la criminalité." *Revue philosophique* 29 (1890): 505–18.

– "Socialisme et criminalité." *Revue philosophique* 16 (1883): 512–20.

– "Théorie organique de la société." *Annales de l'Institut international de sociologie* 4 (1897): 169–341.

– "Les Traits communs de la nature et de l'histoire." *Revue philosophique* 14 (1882): 270–91.

– "Le Type criminel." *Revue philosophique* 19 (1885): 593–627.

Thirard, Jacqueline. "La Fondation de la 'Revue philosophique'." *Revue philosophique* 166, 4 (Oct.-Dec.1976): 401–13.

Thooris, Alfred. "Proportions morphologiques et aptitudes sportives." *L'Ethnographie* n.s., 5 (1922): 25–30.

Thulié, Henri. "La Femme, Education normale." *L'Homme* 2 (1885): 356–61.

– "De la Femme, Essai de sociologie physiologique." *Revue d'anthropologie* 2nd series, 8 (1885): 229–46.
– ed. *1876–1906 L'Ecole d'anthropologie de Paris*. Paris: Félix Alcan, 1907.
Topinard, Paul. "Anthropologie, ethnologie, et ethnographie." *Bulletins de la Société d'anthropologie de Paris* 2nd series, 11 (1876): 199–215.
– "Conclusions et applications de l'anthropologie." *L'Anthropologie* 4 (1893): 657–96.
– "Correspondance." *Revue d'anthropologie* 2nd series, 5 (1882): 748.
– *Eléments d'anthropologie générale*. Paris: A. Delahaye et E. Lecrosnier, 1885.
– "Etude sur la taille." *Revue d'anthropologie* 5 (1876): 34–83.
– *L'Homme dans la nature* [1891]. Paris: Jean-Michel Place, 1990.
– "De la Notion de race en anthropologie." *Revue d'anthropologie* 2nd series, 2 (1879): 589–660.
– Review of Bernhard Blechmann [misprinted as Betchmann], "Ein Beitrag zur Anthropologie der Juden." *Bulletins de la Société d'anthropologie* 3rd series, 6 (1883): 84.
– Review of Girard de Rialle, "Les peuples d'Afrique et d'Amérique." *Revue d'anthropologie* 2nd series, 3 (1880): 673–6.
– Review of Cesar Lombroso, "L'Anthropologie criminelle et ses récents progrès." *L'Anthropologie* 1 (1890): 475–8.
– Review of Alfred Russel Wallace, "Le Darwinisme." *L'Anthropologie* 3 (1892): 326–31.
– Review of August Weismann, "Essai sur l'hérédité et la selection naturelle." *L'Anthropologie* 3 (1892): 331–6.
– *Science et foi, L'Anthropologie et la science sociale*. Paris: Masson, 1900.
– *La Société, L'école, le laboratoire et le musée Broca*. Paris: Typographie G. Chamerot, 1890.
Trautmann-Waller, Céline. *Aux origines d'une science allemande de la culture: Linguistique et psychologie des peuples chez Heymann Steinthal*. Paris: CNRS, 2006.
Tremeau de Rochebrune, Alphonse. "Etude morphologique, physiologique, et ethnographique sur la femme et l'enfant de la race oulove." *Revue d'anthropologie* 2nd series, 6 (1883): 294–8.
Triggs, Mme. O. Lowell [sic]. "La Question nègre au point de vue du Sud." *Revue internationale de sociologie* 6 (1898): 197–9.
Vacher de Lapouge, Georges. "L'Anthropologie et la science politique." *Revue d'anthropologie* 3rd series, 2 (1887): 136–57.
– "Crânes de gentilshommes et crânes de paysans." *L'Anthropologie* 3 (1892): 317–22.
– "Crânes préhistoriques de Larzac." *L'Anthropologie* 2 (1891): 681–95.
– "Le Darwinisme dans la science sociale." *Revue internationale de sociologie* 1 (1893): 414–36.
– "La Dépopulation de la France." *Revue d'anthropologie* 3rd series, 2 (1887): 69–80.

– "L'Hérédité dans la science politique." *Revue d'anthropologie* 3rd series, 3 (1888): 169–95.
– "De l'Inégalité parmi les hommes." *Revue d'anthropologie* 3rd series, 3 (1888): 9–38.
– "Questions aryennes." *Revue d'anthropologie* 3rd series, 4 (1889): 181–93.
– "Sur l'Hérédité." *Revue d'anthropologie* 3rd series, 1 (1886): 512–21.
Valentino, Charles. "L'Evolution de l'intelligence sous le régime des castes." *Revue internationale de sociologie* 15 (1907): 179–217.
Van Gennep, Arnold. Review of Boas, "Anthropology." *Revue des études ethnographiques et sociologiques* 1 (1908): 289–90.
– "La Signification du 1er Congrès d'ethnographie (Neuchâtel 1914)." *Mercure de France* 110 (1914): 322–32.
Vaschide, N[icolas] and Madelaine [sic] Pelletier. *Recherches expérimentales sur les signes physiques de l'intelligence.* Paris: Travail du Laboratoire de Psychologie Expérimentale de l'Ecole des Hautes Etudes [Asile de Villejuif]), 1904.
Vatin, François. *Trois essais sur la genèse de la pensée sociologique.* Paris: La Découverte/M.A.U.S.S., 2005.
Vermeulen, Hans F. "Origins and institutionalization of ethnography and ethnology in Europe and the USA, 1771–1845," 39–59. In Hans Vermeulen and Arturo Alvarez Roldán, eds. *Fieldwork and Footnotes: Studies in the History of European Anthropology.* London: Routledge, 1995.
Verneau, René. "Les Boers et les races de l'Afrique australe." *Revue générale des sciences* 10 (1899): 895–911.
– "Congrès international d'ethnologie et d'ethnographie," *L'Anthropologie* 25 (1914): 367, 370.
– "Discours d'ouverture, cours d'anthropologie." *L'Anthropologie* 21 (1910): 257–79.
– *L'Homme, Races et Coutumes.* Paris: Larousse, 1931.
– *Les Origines de l'humanité.* Paris: Rieder, 1926.
– *Les Races humaines.* Paris: J.-B. Baillière et fils, 1890.
– Review of Adolphe Bloch, "De l'origine des brachycéphales néolithiques de la France." *L'Anthropologie* 12 (1901): 547–8.
– Review of Deniker, "Les Races et les peuples de la terre." *L'Anthropologie* 11 (1900): 762–6.
– Review of Henri Hubert and Marcel Mauss, "Essai sur la nature et la fonction du sacrifice." *L'Anthropologie* 11 (1900): 106–7.
Verrier, Eugène. "Les Aptitudes de la femme et sa craniologie." *Bulletin de la Société d'ethnographie* 2nd series, 6 (1892): 242–9.
– "Ethnographie médicale des peuples de race jaune." *Bulletin de la Société d'ethnographie* 2nd series, 2 (1887): 39–51.
– *De l'Influence des milieux dans la lutte pour l'existence,* 41–67. Paris: Bureau de la Société ethnographique, 1892.

– " De l'Obstétrique comparée dans les races humaines." *L'Homme* 1 (1884): 269–72.
– "Les Races noires de l'Afrique." *Bulletin de la Société d'ethnographie* 2nd, 3 or 13 (1889): 116–28.
– Review of Charles Letourneau, "L'Evolution de la propriété," *Revue orientale et américaine* 2nd series, 4 or 14 (1890): 93.
Vibert, Paul. *La Colonisation pratique et comparée.* 2 vols. Paris: E. Cornély, 1904.
– Discussion of Louis Skarzynski, "Les Peaux-Rouges et les Nègres aux Etats-Unis." *Revue internationale de sociologie* 18 (1910): 355.
Vogt, W. Paul. "Un Durkheimien ambivalent: Célestin Bouglé, 1870–1940." *Revue française de sociologie* 20 (1979): 123–39.
Waller, John C. "Putting Method First: Re-Appraising the Extreme Determinism and Hard Hereditarianism of Sir Francis Galton." *History of Science* 40, 1 (2002): 35–62.
Wartelle, Jean-Claude. "La Société d'Anthropologie de Paris de 1859 à 1920." *Revue d'histoire des sciences humaines* 10 (2004): 125–71.
Weinstein, Leo. *Hippolyte Taine.* New York: Twayne, 1972.
[Weisgerber, Henri.] Review of Fishberg, "Les Juifs: Etude de race et de milieu." *Revue de l'Ecole d'anthropologie* 21 (1911): 407.
Whitelaw, Emma. "Epigenetics: Sins of the Fathers, and their Fathers." *European Journal of Human Genetics* 14 (2006): 131–2.
Wilder, Gary. "Colonial Ethnology and Political Rationality in French West Africa." *History and Anthropology* 14, 3 (2003): 219–52.
– *The French Imperial Nation-State: Negritude and Colonial Humanism between the Two World Wars.* Chicago: University of Chicago Press, 2005.
Williams, Elizabeth A. "Anthropological Institutions in Nineteenth-Century France." *Isis* 76 (1985): 331–48.
– *The Physical and the Moral: Anthropology, Physiology, and Philosophical Medicine in France, 1750–1850.* Cambridge: Cambridge University Press, 1994.
– "The Science of Man: Anthropological Thought and Institutions in Nineteenth-Century France." Ph.D. dissertation, Indiana University, 1983.
Winther, Rasmus. "August Weismann on Germ-Plasm Variation." *Journal of the History of Biology* 34 (2001): 517–55.
Wolf, Theta. *Alfred Binet.* Chicago: University of Chicago Press, 1973.
Worms, René. "Le Concept de la société." *Revue internationale de sociologie* 11 (1903): 81–93.
– Discussion of La Grasserie, "La Race et le milieu comme facteurs de l'évolution sociale." *Revue internationale de sociologie* 10 (1902): 384–5.
– *Organisme et société.* Paris: V. Giard and E. Brière, 1895.
– *Principes biologiques de l'évolution sociale.* Paris: V. Giard and E. Brière, 1910.
– Review of Bouglé, "La Démocratie devant la science." *Revue internationale de sociologie* 15 (1907): 817–18.

– Review of Bouglé, "Les Idées égalitaires." *Revue internationale de sociologie* 8 (1900): 315–17.
– Review of Fouillée, "Psychologie du peuple français," *Revue internationale de sociologie* 6 (1898): 658–9.
– Review of Fouillée, "Tempérament et caractère selon les individus, les sexes et les races." *Revue internationale de sociologie* 3 (1895): 1057–9.
– Review of "La Quinzaine coloniale." *Revue internationale de sociologie* 5 (1897): 238.
– Société de sociologie, 14 déc. 1898. "Les Bases sociales d'après Auguste Comte." *Revue internationale de sociologie* 7 (1899): 48–53.
Yee, Jennifer. "Métissage in France: A Postmodern Fantasy and its Forgotten Precedents." *Modern and Contemporary France* 11, 4 (2003): 411–25.
Zaborowski, Sigismond. "De l'Assimilation des indigènes algériens." *Bulletins de la Société d'anthropologie de Paris* 4th series, 8 (1897): 490–2.
– "Disparité et avenir des races humaines." *Revue mensuelle de l'Ecole d'anthropologie* 3 (1893): 38–9.
– "Ethnographie," 61–5. In Henri Thulié, ed. *1876–1906 L'Ecole d'anthropologie de Paris*. Paris: Félix Alcan, 1907.
– "Hérédité. II. Anthropologie." In André Berthelot, ed. *La Grande Encyclopédie*, 19:1169–70. Paris: H. Lamirault, 1894.
– "Les Méditerranéens et les nègres, la dépopulation et les colonies, les Marocains et L'Europe." *Revue anthropologique* 23 (1913): 149–92.
– Review of Alfred Etiévant, "Choses de l'Algérie." *Bulletins de la Société d'anthropologie de Paris* 8 (1897): 135–40.
Zerilli, Filippo. "Il Dibattito sul meticciato. *Biologico* e *sociale* nell'antropologia francese del primo Novecento." *Archivio per l'antropologia e la etnologia* 125 (1995): 237–73.
– "Etnografia e etnologia al congresso di Arnold van Gennep (Neuchâtel, 1–5 giugno 1914)." *La Ricerca folklorica* 37 (1998): 143–52.
– *Il Lato oscuro dell'etnologia: il contributo dell'antropologica naturalista al processo di istitutzionalizzazione degli studi etnologici in Francia*. Roma: CISU, 1998.
– "Maurice Delafosse, entre science et action," 144–66. In Jean-Loup Amselle and Emmanuelle Sibeud, eds. *Maurice Delafosse, Entre orientalisme et ethnographie*. Paris: Maisonneuve et Larose, 1998.

Index

African-Americans, 62, 65–6, 75, 153–4, 202
Africans, 18–19, 33, 36, 42, 51, 56–8, 60, 68, 76–7, 92–3, 97, 121–2, 138, 147–50, 170, 182, 191, 197, 210; Baule and Liberians, 170–1
Agassiz, Louis, 97
Algeria, 74–80, 176, 178, 211; Kabyles, 75, 78, 178
"Ambiantisme," 25, 40–1
Americans, Native, 91, 93, 117, 149, 173–6, 181, 192
Ammon, Otto, 130
Andrieu, Bernard, 131
L'Année psychologique, 127
L'Année sociologique, 142, 148, 193
Anthony, Raoul, 175, 190
L'Anthropologie, 47, 55, 61, 170, 172–3, 183
anthropometry, crisis of, 6, 66–8, 81–2, 133, 145, 169, 175, 181
anthroposociology, 102, 148, 155, 167, 195, 204
anti-Semitism, 10, 36, 39, 51, 53–4, 67, 154, 194–5, 197–200, 201–4, 206, 210; critique by Manouvrier, 154; by Novicow, 150; by Roberty, 151; by Tarde, 117; by Topinard, 51; and labelling of Jews, 10, 34, 37, 44, 62, 67, 81, 97–8, 152, 176, 178
Arréat, Lucien, 110
Asia and Asians, 10, 15–16, 19–20, 24, 35, 44, 63, 93, 112, 122, 149, 155, 184–5, 195, 197–9, 205–6, 210
Assézat, Jules, 78
associationism and empire, 44, 76–7, 157, 167, 171
Aubin, Joseph Marius Alexis, 21
Auclert, Hubertine, 165, 168
Australians, 34, 49, 55, 58, 60, 63–4, 75, 97, 174–5, 181–2
Avril de Saint-Croix, Ghénia, 165

Bain, Alexander, 88, 126
Balbi, Adrien, 17
Balbiani, Etienne, 126
Barberis, Daniela, 9
Barre, André, 166, 168
Bauer, Arthur, 109, 166
Beaunis, Henry-Etienne, 86, 127
Beauvois, Eugène de, 34
Bérenger-Féraud, Laurent-Jean-Baptiste, 77, 171
Bernard, Claude, 22, 43
Bernard, Félix-Edouard, 113
Bert, Paul, 77, 81
Bertholon, Lucien, 79

Bertillon, Jacques, 135
Bertillon, Louis-Adolphe, 26, 58–9, 207
Bertolini, Mara Meletti, 8, 85
Betts, Raymond, 76
Beuchat, Henri, 172, 177
Binet, Alfred, 9, 68, 86, 126–41, 207; and social circumstances, 136–8
Blanckaert, Claude, xi, 6, 8, 50, 56
blood chemistry, 197
Boas, Franz, 11, 62, 64, 177, 184
Bonneau, Alexandre, 42
Bonnetty, Augustin de, 22
Bordier, Arthur, 78, 81, 197
Boudin, Jean-Christian-Marc, 74
Bouglé, Célestin, 102, 109, 113, 119–21, 125, 148, 175, 208
Boule, Marcellin, 170, 175, 183
Bourgeois, Léon, 22, 133, 185
Bourgoing, Paul-Charles-Amiable, baron de, 20, 41
Bourneville, Désiré Magloire, 69
Boutmy, Emile, 96, 185
Brasseur de Bourbourg, Charles-Etienne, 21
Broca, Paul, 6–7, 11, 23, 25, 27–8, 30, 41, 47–9, 81, 83, 86, 96, 129–30, 178, 181, 183, 190, 196, 201; on hybridity, 19, 49; and phrenology, 32; on women, 69, 72
Brooks, John, 85
Brunhes, Jean, 185–6
Brunschvicg, Cécile, 161
Buffon, Georges-Louis Leclerc de, 17
Buisson, Ferdinand, 132
"Bushmen," 55, 62, 64, 103, 174

Calmette, Emile, 35
Camper, Peter, 52
Candolle, Alphonse de, 92–3, 98, 100, 119
Caro, Elme, 85, 93, 99
Carroy, Jacqueline, xi, 85
Carson, John, 138–9
Cartailhac, Emile, 47, 172
Castaing, Maurice-Alphonse, 24, 26–7, 30–3; on empire, 42–3; on nationalities, 37, 39
cephalometry (head measurement), 127–31, 134–6, 197
Chailley-Bert, Joseph, 81
Chapuis, Elisabeth, 9, 131
Charcot, Jean-Martin, 89, 100, 126
Charencey, Hyacinthe de, 34, 185
Chauffard, Emile, 165
Cheysson, Emile, 161–2, 164, 200
Chinese language, 20, 24, 34
Chinese people, 36, 39–40, 58, 75, 79, 93, 110, 148, 155, 177
Clamadieu, J.-A., 162–3
Clemenceau, Georges, 75
Cloquet, Jules, 184
Cohen, Marcel, 184
Collège de France, 20–1, 23–4, 65, 67, 126, 129, 177
Collège libre des sciences sociales, 151, 165, 177, 185
Collignon, René, 79
Compayré, Gabriel, 110
Comte, Auguste, 9, 84, 141, 161, 163–4
congresses, international: of anthropology, 52, 62n74; colonial expositions, 74; colonial sociology, 76; of International Institute of Sociology, 151; of pedology, 68; on rights of women and feminism, 70, 161. *See also* ethnography and ethnology
Conklin, Alice, 156, 167, 190
Cordier, Henri, 177
Corre, Armand, 77
Cortambert, Eugène de, 42
Cosentini, Francesco, 163

Coste, Adolphe, 150–1, 153, 155
Cousin, Victor, 83, 85
craniology, 24–5, 30, 49–50, 53, 111, 167; and cephalic index, 54, 62, 67, 77, 116, 120, 148
Cuénot, Lucien, 115
Cureau, Adolphe, 80

Daladier, Edouard, 190
Dally, Eugène, 28n44, 29, 71, 97
Daniel, Gaston, 68–9
Darquier de Pellepoix, Louis, 199, 203
Darwin, Charles, 50, 52, 57, 86, 91, 94–5
Darwinism, 73, 120–1, 142, 208
degeneracy, 90, 104, 114, 179, 181
Dejerine, Joseph-Jules, 114
Delafosse, Maurice, 63, 170–2, 176, 184, 190, 192
Delage, Yves, 113
Delbet, Ernest, 163–4, 166, 168
Delvaille, Jules, 118
Demolins, Edmond, 118
Deniker, Joseph, 41, 60, 63–4, 81–2, 130, 177, 182–3, 196, 201
Despine, Prosper, 104
Diamond, Jared, 13, 60
Dias, Nélia, 76
Dissard, Clotilde, 161
Douarche, Léon, 155
Dreyfus affair, 66–7, 119, 125, 128, 149, 151, 154, 167, 169, 192
Drumont, Edouard, 39
Dugas, Ludovic, 115
Duhousset, Louis-Emile, 35
Dumas, Georges, 106
Dumas, Victor, 31
Dumont, Arsène, 155
Duprat, Guillaume-Léonce, 153–5, 167
Durand de Gros, Joseph-Pierre, 59, 207
Durkheim, Emile, 8–9, 11, 30, 102, 109, 118, 120–1, 125, 149, 152, 155, 167–9, 172, 175, 183, 187, 189, 192–3, 207–8; and the biological, 142–5; on gender, 188; and Ribot, 195–7

Ecole d'anthropologie, 10, 23–4, 47, 56, 60, 65, 73, 75, 108, 152, 172, 179, 186–7, 195, 201
Ecole coloniale, 171
Ecole libre des sciences politiques, 185
Ecole pratique des hautes études, 20, 23, 65, 127
Ecole spéciale des langues orientales, 20–1, 23, 177
Edison, Paul, 33, 37, 40
Edwards, William Frédéric, 18
Espinas, Alfred, 87, 101, 119–20, 148, 208
ethnic crossing, commission for study of, 182
"ethnie," 196, 202, 210
L'Ethnographie, 184–5
ethnography and ethnology, 17, 21, 25; debate on definitions, 27–9; international congresses, 22–3, 27, 31, 36, 62–3, 178
eugenics, 5, 40, 54–5, 91, 148, 180, 199
evolutionary psychology, 11

Faculty of Letters, Paris (Sorbonne), 84, 86, 127, 209
Faculty of Medicine, Paris, 23, 77, 196, 199
Faidherbe, Louis, 63, 77
Fancher, Raymond, 13
feminism, 12, 71–2, 132, 146, 158–68, 208–9; anti-feminism in Martial, 200. *See also* gender roles

Ferri, Enrico, 117
Ferry, Jules, 74
Finot, Jean, 122, 154, 160
Firmin, Anténor, 67, 113, 150, 163
Fishberg, Maurice, 67
Flourens, Jean-Pierre, 32
Foa, Edouard, 80
Foley, Antoine-Edouard, 34
Fouillée, Alfred, 96, 108, 115–16, 120, 123, 125, 147, 151, 208
Franck, Adolphe, 98–9
Frazer, James, 179
Freeden, Michael, 5
Fuegians, 55, 63, 174

Gall, Franz Joseph, 32, 94
Galton, Francis, 5, 54, 90–1, 95, 98, 101, 106, 119, 126, 128, 130–1, 155, 179–80, 182, 207; on Ribot, 99–100; on twins, 12
Gambetta, Léon, 49, 163
Garcin de Tassy, Joseph-Héliodore-Sagesse-Vertu, 21
Garnier, Adolphe, 32
Gates, Reginald Ruggles, 197
Gaucher, Fernand, 155
gender roles, 9, 12, 14–16, 32, 36, 46–7, 69–73, 111, 122–4, 144–6, 158–68, 188, 200, 208–10
Gennep, Arnold van, 45, 62–3, 170, 176–8, 184–5, 189, 192, 209
Geoffroy Saint-Hilaire, Isidore, 32–3
Geslin, Charles, 43
Girard de Rialle, Julien, 28
Gissis, Snait, 9
Gliddon, George, 98
Gobineau, Joseph-Arthur de and Gobinian, 29, 49, 50n16, 55, 97, 120, 147, 154, 180–1, 196–7, 210
Golberg, Mécislas, 149–50
Gould, Stephen Jay, 131
Grimanelli, Paul, 160–1, 166, 168
Guieyesse, Paul, 185
Guyau, Jean-Marie, 108, 110, 151
Guyot, Yves, 74–6, 82

Haiti and Haitians, 40, 42, 44, 55, 67, 98, 147, 150, 154, 163, 206
Hamy, Ernest-Théodore, 25, 30, 47, 60–1, 76, 170
Harmand, Jules, 80, 156, 171
Hecht, Jennifer, 6
Henri, Victor, 127
hereditary disease, 88–90, 101, 196
heredity, theories of, 53
Hering, Ewald, 88
Hervé, Georges, 25, 29, 39, 52, 191, 201; on racial mixing, 181–2, 201
Hervey de Saint-Denis, Marie-Jean-Léon, marquis de, 34–5
Hirszfeld, Ludwik and Hanna, 195
L'Homme, 29, 35, 47, 65, 70
Hovelacque, Abel, 38, 52, 56, 75
Hoyt, David, 169
Hubert, Henri, 67, 172

immigration, into France, 15, 33, 36, 195, 197–9, 211
imperial expansion of France, 16, 40–4, 205, 209; anthropologists on, 47, 73–81; ethnographers' resolutions on, 42–4; sociologists on, 156–8, 167
Indochina, 20, 22, 42, 74–5, 77, 80, 156, 171, 177, 183, 185, 191
Institut de France, 20–1
Institut ethnographique international, 63, 176–7, 192
Institut d'ethnologie, 175, 178, 187–91, 193, 209
Institut français d'anthropologie, 10, 63, 175, 183–4, 201, 209

Institut international de sociologie, 65–6, 116–17, 143, 147, 149, 151, 160, 167
intelligence testing, 132–6
Isidor, Lazare, 40
Issaurat, Cyprien, 26, 38

Jacoby, Paul, 95
Janet, Paul, 85, 99
Jankélévitch, Samuel, 120–2, 125, 148, 208
Japanese language, 20, 185
Japanese people, 20, 22, 39, 93, 137, 148, 174, 201
Jastrow, Joseph, 128
Joly, Henri, 105, 122, 125, 154, 180, 208
Jomard, Edme, 20–1
Joran, Théodore, 164–6, 168

Kellès-Krauz, Casimir de, 157
Kollar, Adam, 17
Kossovitch, Nicholas, 195
Kovalewsky, Maxime, 151, 153

Labarthe, Jean-François-Charles, dit de, 24, 26, 31, 34–5; on empire, 42–3; on nationalities, 37
Lacombe, Paul, 109
Lagneau, Gustave, 59, 207
La Grasserie, Raoul de, 151–4, 161, 164, 166, 179
Lamarck, Jean-Baptiste, 47–8, 94, 113, 115
Lamarckianism, 7, 47–8, 52, 56, 61, 66, 81, 87, 94, 106, 113–15, 119, 121–2, 124, 160, 179, 196–7, 206–8
Lambert, Alfred, 153–5, 157, 159–60, 163, 167
Lanessan, Jean-Marie-Antoine, 77, 185
Lapicque, Louis, 130, 182–4
Lapie, Paul, 124
Lazarus, Moritz, 96
Le Bon, Gustave, 4, 32, 48, 53–4, 66, 81–2, 96, 101, 108, 111–12, 116, 120, 123, 125, 143, 150, 152, 154, 156, 164, 187, 201, 206; on empire, 76; on gender, 69–71
Lebovics, Herman, 33, 37, 190
Le Dantec, Félix, 114–15, 119, 121, 124, 208
Le Pen, Jean-Marie, 211
Le Play, Frédéric, 118, 161, 185, 200
Leroy-Beaulieu, Paul, 157
Lesseps, Baron Jules de, 22
Letourneau, Charles, 27, 58, 68, 72, 81–2, 97, 100, 108, 112, 116, 120, 123–5, 166, 171–2, 179, 206; on gender, 69–70, 123, 160; on race and milieu, 56–7
Lévi-Strauss, Claude, 190
Lévy-Bruhl, Lucien, 173, 183–4, 190
Lewes, George, 91
Lewontin, Richard, 13
Limousin, Charles-Mathieu, 149, 151–3, 156, 162, 166–8
Linnaeus, 17
Lombroso, Cesare, 65, 117, 128, 179–80
Loria, Achille, 163
Lourbet, Jacques, 123
Lucas, Prosper, 89, 91, 94

MacAuliffe, Léon, 185–6
Madier de Montjau, Edouard, 36
Maguerie, Jeanne de, 157, 161, 164, 166
Malinowski, Bronislaw, 169
Manouvrier, Léonce, 29–30, 44, 68, 82, 113, 116–17, 120, 123, 125, 129, 133, 135, 140, 145, 150,

152–3, 163, 167, 179–81, 192, 206–7; on brain size and milieu, 64–7; on feminism, 12, 70–2, 159–60, 166; on sociology, 27
Marey, Etienne-Jules, 128
Marie, Auguste, 185–6
Marin, Louis, 185–7, 190–1, 195, 201
Marion, Henri, 112
Martial, Lydie (Anna Carnaud), 162, 164
Martial, René, 195–201, 204, 209–11
Matiegka, Heinrich, 131
Maudsley, Henry, 95, 104
Mauger, Gérard, 203
Maunier, René, 165–6
Maury, Louis-Ferdinand-Alfred, 20
Mauss, Marcel, 167, 172–3, 178, 183–4, 189–91, 202
Mauxion, Marcel, 112
Mendelianism, 115, 182, 184, 197
mésologie (science of milieu), 26, 58
Mill, John Stuart, 126, 160, 163–4
Minoret, Eugène, 34, 43
Misme, Jane, 166
Moll-Weiss, Augusta, 164, 166
Monin, Hippolyte, 162, 166
monogenist-polygenist debate, 73–7
Montandon, George, 195–6, 201–4, 209–10
Montesquieu, Charles-Louis de Secondat, baron de, 118
Moreau de Tours, Jacques, 89–90, 95
Morel, Bénédict-Augustin, 90, 95, 101
Morgan, Jacques de, 177
Morocco, 80
Mortillet, Adrien de, 63, 65, 172–3, 175
Mortillet, Gabriel de, 35, 65, 74
Moussy, Martin de, 97
Mucchielli, Laurent, xi, 8, 85, 106, 108
Müller, Max, 105
Musée d'ethnographie du Trocadéro, 21, 61, 173, 175, 185
Musée de l'homme, 173
Muséum national d'histoire naturelle, 25–6, 57, 60–3, 81, 130, 175, 182, 185, 201, 206

Nadaillac, Jean-François-Albert du Pouget, marquis de, 60
National Front, 211
nationalities, 36–8; and Alsace, 38–9; and national character, 101; and Romania, 38
nature-nurture debate, 3, 10–16, 92–3, 211–12
Nicolas, Serge, 8, 85, 98
Nott, Josiah, 98
Novicow, Jacques, 119–21, 124–5, 150–1, 153, 155, 159, 164, 166–7, 208

Offen, Karen, 146, 158
Oppert, Jules, 27
Orbigny, Alcide d', 97
organic model of society, 117–20, 142, 156

Paligot, Carole Reynaud, xi, 5, 8, 77, 80, 83
Papillault, Georges, 65, 152, 167, 179–82, 191
Papillon, Fernand, 99
Parodi, Dominique, 119
Paul, Diane, 5
Paulhan, Frédéric, 110–12
Pearson, Karl, 131–2
Pelletier, Madeleine, 131–2, 162, 164–6, 168, 209
Périer, Joanny-André-Napoléon, 97
Perrier, Edmond, 185
Philippe, Léon, 156
phrenology, 32, 44, 65

physiognomy, 68, 134
Piéron, Henri, 63, 105, 127, 175
Pinker, Steven, 10–13, 94
Plas, Régine, 85
Polynesians, 34, 64, 97
Ponsinet, L., 163
Porter, W. Townshend, 130
Pouchet, Georges, 74
Pournin, Marcel, 157
Pressensé, Francis de, 67
Pruner-Bey, Franz, 58
psychological heredity, 93–102
psychology, experimental, 86, 127
psychology of peoples, 96–7

Quatrefages, Armand de, 27, 29, 32, 44, 57–8, 62–3, 76, 81, 98, 112–13, 206

Rabaud, Etienne, 183–4
races, 18–20, 29–31, 33–4, 53; Aryan theory, 21, 38, 55–6, 58, 109, 154, 175–6, 197, 202; in Broca, 49–50; in Deniker, 64; in Fouillée, 115–16; in Galton, 92; in Hovelacque and Hervé, 52; and imperialism, 73–81; in Le Bon, 54; in Letourneau, 56–7; in Manouvrier, 65–8, 150; in Martial, 195–200; in Montandon, 201–4; in Papillault, 179–82; in Quatrefages, 57–8; in Ribot, 97, 104; in Rivet, 173–6; in Spencer, 87; in Tarde, 116–17; in Topinard, 51; in Vacher de Lapouge, 54–5; in Verneau, 61; in Worms, 147–8; in Worms group, 149–55
Radcliffe-Brown, Alfred, 169
Regnault, Félix, 196
Reinach, Salomon, 172, 183
Renan, Ernest, 21, 38
Renouvier, Charles, 99
republican universalism, 4, 144, 146–7, 167–8, 210
Retzius, Anders, 33, 77
Réville, Albert, 103
Revue d'anthropologie, 47, 54
Revue d'ethnographie (1882–89), 30, 47, 61
Revue d'ethnographie et de sociologie, 176
Revue internationale de sociologie, 9, 101, 117, 143–4, 148, 153–5, 167
Revue orientale et américaine, 41
Revue philosophique, 8, 85, 102, 108–26, 141, 152, 154, 207; anti-Durkheimian backlash, 120–2; and Fouillée, Tarde, and Worms, 115–18; and gender roles, 122–4; Le Bon and Letourneau, 111–12; and rise of sociology, 118–20; on Taine and Spencer, 109–11; and Weismann, 113–14
Ribot, Théodule, 8, 15, 84–107, 109, 116, 118, 121, 124–6, 131–3, 140–1, 143, 147, 155, 196, 207–8; autonomy of psychology in, 84–6; concessions on inheritance, 101–2; and Galton, 90–3; reviews of thesis, 98–101. *See also* psychological heredity and psychology of peoples
Richard, Gaston, 109, 112, 120, 123, 125, 147, 208
Richet, Charles, 48, 86, 100, 129
Rivet, Paul, 10, 170, 173–6, 181–4, 187, 190–2, 195, 201
Roberty, Eugène de, 151, 167
Rosny, Léon de, 16, 20, 22, 24, 26–7, 33, 35, 41, 178, 185, 187, 192, 205; on empire, 42–3; on equality, 31; on inequality, 39–40; on nationalities, 37
Rousson, Mme, 134
Royer, Clémence, 27, 34–6, 43, 50, 53, 71–2, 82, 122, 209
Ruskin, John, 197

Sabatier, Camille, 77–8
Sanson, André, 79
Saussure, Léopold de, 76, 157, 170
Schlözer, August Ludwig, 17
Schoebel, Charles, 22, 34, 43
Schrader, Franz (Jean-François), 59–60, 80, 82, 152–3, 167, 207
scientific materialists, 23, 25, 50n18, 56, 65, 169
Scott, Joan Wallach, 146
Seignobos, Charles, 111
Silbermann, Joseph B., 34, 43
Simon, Théodore, 129–30, 139
Société africaine de France, 35
Société asiatique, 19–20
Société d'anthropologie de Paris, 6–7, 10, 46, 61, 63, 73, 172, 174–6, 179–83, 191; membership, 24, 86
Société d'ethnographie (post-1913), 10, 184–7, 192
Société d'ethnographie américaine et orientale, 6, 19, 25, 177; membership, 24
Société française de philosophie, 126
Société libre pour l'étude psychologique de l'enfant, 132–3, 135
Société de psychologie, 65
Société de psychologie physiologique, 100, 126
Société de sociologie (1872–4), 9, 58
Société de sociologie de Paris (post-1895), 108–9, 116–17, 122, 143, 146, 149, 152–3, 167, 179; on gender, 158–66
Société des américanistes, 19–20, 174
Société ethnologique de Paris, 4, 18–20
Soustelle, Jacques, 176
Spencer, Herbert, 9, 86–7, 90, 95, 103, 108, 110–11, 124–6, 151, 207–8
Spurzheim, Johann Caspar, 32
Steinthal, Heymann, 96
Stepan, Nancy, 145
Surkis, Judith, 158
Sylvain, Benito, 40
Sylvain, Georges, 154

Taguieff, Pierre, 198
Taine, Hippolyte, 84–7, 96, 98, 108–11, 124–6, 151, 196, 208
Tarde, Gabriel, 108, 115–18, 124, 143, 152–3, 161, 180, 208
Taylor, Harriet, 164
Thierry, Amédée and Augustin, 18
Thooris, Alfred, 186
Thulié, Henri, 56, 70
Topinard, Paul, 11, 28–9, 47, 50–1, 53–4, 64, 71, 86, 130, 201
Triggs, Mrs Oscar Lovell, 154
Tylor, Edward Burnett, 103, 105, 179

Vacher de Lapouge, Claude, 199, 204
Vacher de Lapouge, Georges, 81, 102, 113, 130, 148, 150, 154, 180, 206–7, 210; and "ethnic," 152; and extreme hereditarianism, 53–5; and Manouvrier, 66–7; and Martial, 195–7, 199; and Montandon, 201, 204; and *Revue internationale de sociologie*, 167
Valentino, Charles, 156
Vallat, Xavier, 203
Vallois, Henri, 175, 191
Vaschide. Nicolas 129, 131–2
Venn, John, 130
Verneau, René, 60–2, 67, 81, 169, 172–3, 175, 178, 183–4, 191
Verrier, Eugène, 25, 27, 35–6, 42–3
Vibert, Paul, 154, 157, 162, 166–7
Vichy regime, 10, 54, 153, 176, 194–205, 210
Vinson, Julien, 172

Vlangali-Handjéri, Prince Michael, 20
Vogt, Carl, 67, 88
Vries, Hugo de, 114

Waitz, Theodor, 96, 98
Wallace, Alfred Russel, 103
Wartelle, Jean, 7
Weber, Marianne, 188
Weisgerber, Henri, 67, 182
Weismann, August, 94, 101–3, 113–14, 121, 124, 208
Williams, Elizabeth, 6–7, 23, 37, 61
Worms, René, 9, 30, 65–6, 101–2, 108, 115–16, 125, 141–5, 153–5, 164, 166–7, 208; on race, 117–18; on race and colonization, 147–8, 154, 156

Zaborowski, Moindron-Sigismond, 28–9, 53–4, 79–82, 181–2
Zerilli, Filippo, 10, 191